Solvent Mixtures

Solvent Mixtures

Properties and Selective Solvation

Yizhak Marcus

The Hebrew University of Jerusalem
Jerusalem, Israel

MARCEL

DEKKER

MARCEL DEKKER, INC.   NEW YORK · BASEL

ISBN: 0-8247-0837-7

This book is printed on acid-free paper.

Headquarters
Marcel Dekker, Inc.
270 Madison Avenue, New York, NY 10016
tel: 212-696-9000; fax: 212-685-4540

Eastern Hemisphere Distribution
Marcel Dekker AG
Hutgasse 4, Postfach 812, CH-4001 Basel, Switzerland
tel: 41-61-260-6300; fax: 41-61-260-6333

World Wide Web
http://www.dekker.com

The publisher offers discounts on this book when ordered in bulk quantities. For more information, write to Special Sales/Professional Marketing at the headquarters address above.

Current printing (last digit):
10 9 8 7 6 5 4 3 2 1

PRINTED IN THE UNITED STATES OF AMERICA

Preface

Much of chemistry is carried out in solution, and often mixed solvents rather than single solvents are used for this purpose. In the solutions, the solutes, whether nonelectrolyte molecules or ions, are solvated. The extent and strength of this solvation affect their solubility, reactivity, and other properties. In mixed solvents, the mutual interactions of the solvents have to be taken into account in assessing the solvation of the solutes. It is therefore of interest to learn as much as possible about the properties of the solvent mixtures, first in the absence of any solutes and then in their presence. Contrary to solvation in single solvents, in solvent mixtures the phenomenon of preferential solvation may take place. This adds to the difficulty in interpreting experimental results, but, on the other hand, is of interest in itself. Preferential solvation may occur in a binary solvent mixture in the absence of solutes, in which molecules of a solvent may interact preferentially with those of its own kind or with those of another kind. When solute molecules or ions are present, these may also be preferentially solvated by a certain component of the solvent mixture. Therefore, the use of mixed solvents entails knowing their properties as functions of the composition as well as an understanding of any preferential solvation that takes place. It is the purpose of this book to provide a solid basis for this knowledge and understanding.

This book is therefore a fitting sequel to a previous book of mine: *The Properties of Solvents* (Wiley, 1998), which presented in annotated tabular form, the physical, thermodynamic, and chemical properties of some 250 single solvents. It is obviously impossible to present properties of all the combinations of chemically stable and mutually miscible mixtures of these many solvents. A drastic selection of systems to be reported has to be made, from the point of view of the availability of data and the common employment of the mixtures in practice. More than 40 years ago, Jean Timmermans attempted to report *The Physico-chemical Constants of Binary Systems in Concentrated Solutions* (Interscience, 1959–60), pertaining (in Volumes 1 and 2) to a large number of binary solvent mixtures and a sizable number of properties. Most of these data are still useful, but of course have been supplemented—and in some cases replaced—by more recent data. The present book is less exhaustive with respect to the number of solvent mixtures treated, but offers in addition to the data a sufficiently deep discussion of their significance. William E. Acree, Jr., in *Thermodynamic Properties of Nonelectrolyte Solutions* (Academic, 1984), has more recently dealt with some of the points discussed here but reported only few data concerning binary solvent mixtures. The previous gap in knowledge and understanding of the properties of solvent mixtures and of the solutions in them should therefore be narrowed by the present book.

The rationale for my undertaking the writing of books in the field of solution chemistry was partly presented in the preface of *The Properties of Solvents*, taken in part from my earlier book *Ion Solvation* (Wiley, 1986). Gaps in knowledge and understanding revealed by the writing of books could—albeit to only a minute extent—be filled by my own research. However, in view of the intrinsic as well as practical importance of the subject, the scientific community is urged to do more in this respect. For this purpose, indeed, the tables of data contain blank rows where data are missing, suggesting that these should be filled in. The comparison of data, presented here in a uniform manner to facilitate comparison, should go a long way toward aiding comprehension of the behavior of the systems already studied and extrapolation of the insights and knowledge to many other systems.

The data presented here in the extensive tables are from secondary sources as far as available, since such data ought to have already been critically evaluated and selected by the authors of such sources. These data have been supplemented by data from recent primary sources in research journals. Access to these was through abstracts, electronic journals, and the Web of Science up to and including 2001. I am responsible for the choices made and will be grateful for readers' reports of errors, oversights, and further useful data for the systems discussed. The rationale for the selection of these systems is given in Chapter 1. On the other hand, the discussion of the physical chemistry of the mixed solvent systems, the structures, thermodynamics, and the physical and chemical properties of these systems is independent of the data actually available. This discussion points to the significance of the data as well as the reliability and accuracy with which these can be obtained. The discussion is illustrated by results for certain selected systems, but naturally is not comprehensive in this respect.

I am grateful to The Hebrew University of Jerusalem for its policy of permitting its retired professors—of whom I am now one—to retain facilities for the continuation of their scientific work. It also provides a fairly generous allocation of funds for doing so, as long as the retired professors are able to do useful work. This book was conceived before my retirement but I have written it while being an emeritus professor, a status that I keep on enjoying.

Yizhak Marcus

Contents

Solvent Abbreviations

These abbreviations are used mainly in tables, where the space is too narrow for the full name, but also sparingly in the text, if names are often repeated.

t-BuOH	2-methyl-2-propanol, tertiary butanol
DMA	*N,N*-dimethylacetamide
DMF	*N,N*-dimethylformamide
DMSO	dimethylsulfoxide
EtOH	ethanol
FA	formamide
HMPT	hexamethyl phosphoric triamide
MeCN	acetonitrile, cyanomethane, ethanenitrile
Me$_2$CO	acetone, 2-propanone
MeNO$_2$	nitromethane
MeOH	methanol
NMA	*N*-methylacetamide
NMF	*N*-methylformamide
NMPy	*N*-methylpyrrolidin-2-one
PC	propylene carbonate
i-PrOH	2-propanol
n-PrOH	1-propanol
THF	tetrahydrofuran

Symbols

Latin Letters

A, B, C...	components in a solvent mixture
A	molar Helmholtz energy
A	surface area of liquid
A	(integrated) absorbance of light (of an infrared absorption band)
AN	Mayer–Gutmann acceptor number
a	activity
a	atomic (radiation) scattering intensity
a	molar absorbance
B	second virial coefficient
B_K	Koppel–Lewis basicity parameter
C_P	heat capacity at constant pressure
C_V	heat capacity at constant volume
c	molarity
c	parameter of the simplified UNIQUAC expression

c_j	coefficient of the Redlich to Kister-type heat capacity expression, Eq. (2.34)
D	homogeneous distribution factor in COPS method (Section 5.2.1)
DN	Gutmann donor number
d	density
E^C	cohesive energy
$E_T(30)$	Dimroth–Reichardt polarity parameter
E_T^N	normalized Dimroth–Reichardt polarity parameter
e	interaction energy
F	Faraday constant, $96485\ \mathrm{J\ mole^{-1}}$
f	activity coefficient (rational scale)
f	fraction of specified species
G	molar Gibbs energy
G_{ij}	Kirkwood–Buff integral, Eq. (4.24)
g_j	coefficients of Gibbs energy or chemical potential, Eqs. (2.14, 15, and 17).
g	parameter of the NRTL expression, Eq. (2.26)
g	Kirkwood dipole orientation correlation parameter
g_{ij}	pair correlation function for finding a molecule i given the molecule j, Eq. (3.2)
H	molar enthalpy
ΔH_r	molar enthalpy change for defined reaction r
h	partial molar enthalpy
h_j	coefficient of the enthalpy expression, Eq. (2.31)
I	intensity of absorbed or diffracted radiation
i	electric current
K_r	equilibrium constant of defined equilibrium r
k	suffix of Margules expression
k_B	Boltzmann's constant, $1.3805 \times 10^{-23}\ \mathrm{J\ K^{-1}}$
L	total number of nearest neighbors, Eq. (4.12)
L_D	Debye correlation length

l	relative partial molar enthalpy
M	molar mass
m	molality
m	mass of solvent transported
N	number of molecules
N	solvation number
N_{Av}	Avogadro's number, 6.02214×10^{23} mol^{-1}
N_Z	number of atoms of atomic number Z
n	numbers of moles (amount of substance)
n_D	refractive index (at sodium D-line)
P	pressure
P	molar polarization, Eq. (2.7)
p	vapor pressure
p	solvent sorting parameter, Eq. (4.3)
Q	derivative of the activity coefficient or excess Gibbs energy, Eq. (4.26)
Q	distribution between solvated and complexed species, Eq. (5.17)
R	gas constant, 8.3143 J K^{-1} mol^{-1}
R_D	molar refraction [at sodium D-line, Eq. (2.9)]
R_{cor}	extent of the correlation region, Section 4.2
q	surface area parameter in the UNIQUAC expression, Eqs. (2.27 and 2.28)
q	partial charge on an atom or molecule
r	distance from the center of a molecule
r	volume parameter in the UNIQUAC expression, Eqs. (2.27 and 2.28)
S	molar entropy
S	structure factor, Eq. (3.1)
S	van der Waals surface area
s	angle- and wavelength-dependent variable in diffraction measurements

s_j	coefficient of the Redlich–Kister-type entropy expression
T	absolute temperature
t	saturation factor in COPS method (Section 5.2.1)
t	time
U	molar internal energy
U	potential of mean force, Eq. (3.7)
u	pair-wise interaction potential energy, Eq. (3.8)
u	speed of ultrasound
V	molar volume
v	partial molar volume
v_j	coefficient of the Redlich–Kister-type volume expression, Eq. (2.37)
W	weight
w	weight or mass fraction, Eq. (1.1)
X	a generalized property of a component or mixture, Eq. (4.33, 5.11)
x	mole fraction, Eq. (1.2)
Y	a generalized (commonly thermodynamic) property of a component or mixture
y	mole fraction in the vapor phase (or a special use in Eq. (4.35))
y	packing fraction in a liquid
y	the contribution of pair interaction to the generalized property Y
Z	compressibility factor
Z	Kosower–Lewis-acidity parameter
Z	atomic number of an element, section 3.1
Z	lattice parameter in the QLQC model, Section 4.2
z	charge number of ion (taken algebraically)

Greek Letters

α	parameter of the NRTL expression, Eq. (2.26)
α	Kamlet–Taft Lewis-acidity (hydrogen bond donation) parameter
α_P	isobaric expansivity
α_{Pj}	coefficient of the Redlich–Kister-type expansivity expression, Eq. (2.40)
β	Kamlet–Taft Lewis-basicity parameter
Γ	surface excess concentration
γ	Ostwald coefficient of gas solubility
Δ	difference in solvent transported with ions, Eq. (5.72)
δ	(Hildebrand) solubility parameter
δ	NMR chemical shift, Section 3.3
δ_x	preferential solvation parameter, Eq. (4.2)
ε	relative permittivity
ε_0	permittivity of vacuum, $8.8542 \times 10^{-12}\ \mathrm{C^2 J^{-1} m^{-1}}$
ε	depth of the potential well in the interaction between molecules
η	(dynamic) viscosity
θ	surface fraction pertaining to the UNIQUAC expression
θ	angle at which radiation is diffracted, Section 3.1
κ_S	adiabatic compressibility
κ_T	isothermal compressibility
κ_{Tj}	coefficient of the Redlich–Kister-type compressibility expression, Eq. (2.41)
κ_{iS}	affinity of solute S to solvent i in COPS method (Section 5.21)
Λ	parameter in the Wilson expression, Eq. (2.23)
λ	wavelength of radiation
μ	chemical potential
μ	dipole moment

ν	kinematic viscosity
ν	wavenumber
π^*	Kamlet–Taft polarity/polarizability parameter
ρ	number density (per 1 cm^3)
σ	molecular diameter
σ	surface tension
τ	parameter of the NRTL expression, Eq. (2.26)
ξ	Orenstein–Zernicke correlation length, Section 3.1
ϕ	volume fraction (ignoring volume change on mixing), Eqs. (1.6 and 2.27)
χ	generalized interaction parameter of regular mixtures
χ	fraction complexed in COPS method (Section 5.2.1)

Subscripts, Superscripts, and Phase Designations

Subscripts

AP	autoprotolysis
i,j,k	of component i, j, k (= A, B, etc.)
j	numerical index of coefficient
C	critical
p_C	pseudo-critical
t_C	true critical
cor	correlation region around a molecule
d	dimer
in	internal
P	at constant pressure
PS	of preferential solvation
S	at constant entropy (adiabatic)
S	a cosolvent of water
S	solute
T	at constant temperature
t	of transfer

vdW	van der Waals (intrinsic volume)
X	McGowan (intrinsic volume)
mix	of mixing
V	of vaporization
W	water

Superscripts

E	excess property
id	ideal
L	local, Eq. (4.1)
sat	of the saturated solution
o	of neat component; standard molar quantity
∞	at infinite dilution

Phase Designations

(i.g.)	of ideal gas
(l)	of liquid phase

1

Introduction

1.1 SOLVENT MIXTURES

Much of chemistry is carried out in solution, hence solvents are a necessary and widely employed requisite in chemical processes, both in industry and in the laboratory. For the purpose of this book, a *solvent* is a substance that is liquid at the temperature of application, which other substances can be dissolved in or mixed with to yield a homogeneous isotropic liquid [1]. Water is, of course, the most widely used solvent because of its availability, low cost, nontoxicity, and safety as well as its ability to dissolve a great variety of substances, including electrolytes and polar organic substances. It then permits their study by electrochemical, spectroscopic, and other methods. Certain substances are more soluble in polar organic solvents, which also have other properties that make them useful and desirable solvents. On the other hand, substances of low polarity that are essentially insoluble in water are often readily dissolved in nonpolar organic solvents. In many of these applications the solvents are employed in substantially pure form, as so-called neat solvents, and may have various impurities removed before use [2].

For certain applications, however, neat solvents fall short of the mark as far as their dissolving power or other properties are concerned. It is then expedient to use *solvent mixtures,* which may range from *binary mixtures* involving two solvents to *ternary* (three solvents) or even higher *multicomponent mixtures.* Some of the admixed substances may not be liquids at the temperature of application, hence mixtures involving them are not properly called solvent mixtures. For instance, mixtures of water and urea (aqueous solutions of urea) have been widely employed to study various aspects of hydrogen bonding of the mixtures themselves or of substances dissolved in them. Although such mixtures are similar in many respects to certain aqueous solvent mixtures, say water mixed with formamide, they are not discussed in this book. Then again, the presence of a small amount of one liquid substance in a large excess of another, at mole ratios or fractions of, say, 1:100, may have far-reaching effects on the properties of the major component. Such dilute solutions are proper subjects for study and lead to important insight concerning their behavior but are not ordinarily called "solvent mixtures," hence are also outside the scope of this book. Most of the book is devoted to binary solvent mixtures, in which both components are present in substantial amounts. Still, certain generalizations to higher, multicomponent, solvent mixtures are within the scope of the book and are discussed.

Why, then, are solvent mixtures used in the chemical industry and in laboratories? The foremost reason is for enhancement of the solubility of substances that have too low solubility in neat solvents, i.e., for their *solubilization.* The components of the solvent mixture may interact with different parts of the intended solute and thus have a synergistic effect on the solubility. This aspect is of wide use in the pharmaceutical industry. Consider, as an imaginary example, an amphiphilic solute with a polar nonionic head, such as hexakis(ethylene oxide), and a long hydrocarbon tail, such as hexadecyl. Or consider an ionic solute, such as calcium stearate, where the calcium cation and the carboxylate group of the anion constitute a polar head and the heptadecyl chains the nonpolar tails. Such a solute may not be sufficiently soluble in either neat N,N-dimethylformamide (DMF) or p-xylene for the envisaged purpose. However, the polar component of the mixed solvent (DMF) will solvate the polar or ionic head of the solute and the nonpolar component (p-xylene) will solvate the long tail. Thus, such a mixed solvent may be suitable for the solubilization of these kinds of solutes.

In other cases, the components may confer on the mixture physical properties that enhance solubility, apart from the specific solvation of parts of the solute. For example, higher relative permittivity can be useful if ionic dissociation can increase the solubility. In still other cases the mixed solvent may have *improved physical properties* compared with its neat components, e.g., with respect to density, viscosity, vapor pressure, and the freezing or the boiling temperature. Water, of course, freezes at $0°C$ at ambient pressure and dimethylsulfoxide (DMSO) freezes at $18.5°C$. However, a mixture containing 34 mole% water and 66 mole% DMSO forms a eutectic freezing at $-78.6°C$. Therefore, mixtures of these components (water and d_6-DMSO) at suitable compositions can be cooled down to quite low temperatures. These may be sufficient to slow down the exchange between the protons of protic solutes and water, enough that they show separate nuclear magnetic resonance (NMR) signals.

As another example, tri-*n*-butylphosphate (TBP) is widely used in the nuclear industry for nuclear fuel reprocessing, for the extraction, separation, and purification of uranium and plutonium. Neat TBP when in contact with water or a dilute aqueous solution hydrates to a ratio of 1.05 ± 0.01 H_2O to TBP, and this liquid mixture has a viscosity of 3.39 mPa·sec, some 20% higher than that of neat TBP itself, which has a density of 0.973 g cm^{-3}. Dilution of the TBP with dodecane (or kerosene with similar properties, as practiced in the industry) to a nominal concentration of 19% by volume not only reduces the water content of the organic phase produced but also reduces the viscosity to near that of dodecane itself, 1.38 mPa·sec, and the density to only slightly above that of dodecane, 0.754 g cm^{-3}. Such mixtures are preferred by the industry for solvent extraction processes due to the improved hydrodynamic properties and the lower density of the organic phase, facilitating phase disengagement.

Now, the kerosene mentioned in this example is itself a mixed solvent. The "odorless kerosene" preferred for the dilution of the TBP is a mixture of aliphatic straight-chain saturated hydrocarbons with a boiling range of 195 to 254°C, corresponding to a mixture of undecane, dodecane, tridecane, and tetradecane. Other *industrial solvents* consisting of hydrocarbons may contain alicyclic and aromatic hydrocarbons and are rarely neat solvents or even binary mixtures. Since benzene became disused because of its carcinogenic properties, mixtures of xylenes have become popular aromatic solvents, the commonly used mixtures consisting of $\sim 40\%$ *m*-xylene, $\geq 20\%$ *o*-xylene, $\geq 20\%$ *p*-xylene, and some ethylbenzene. Other solvents commonly used as isomeric mixtures are decalin, *cis*- and *trans*-decahydronaphthalene, and amyl alcohol (refined fusel oil), obtained from fermentation, 85% 3-methyl-1-butanol and 15%

l-2-methyl-1-butanol. Technical grade chloroform usually contains 1% ethanol as a stabilizer against oxidation. Nonylnaphthalenesulfonic acid, used as a liquid cation exchanger, is a mixtures of isomers, both regarding the nonyl group (which may be trimethylhexyl or tetramethylpentyl, at various attachments of the methyl groups) and regarding the position of the nonyl group in the naphthalene residue. For many purposes these and other mixed solvents, obtained as technical grade chemicals, can be used with the advantage of ready low-cost availability in bulk, with no detrimental effects on their utilization. The components of such mixed solvents may have the same functional groups and differ only in their position on an aromatic ring or in the alkyl chain attached to the groups. The chemical properties of the components should then be similar. Although their physical properties, such as vapor pressures and molar volumes, differ, they dissolve and solvate various solutes as if they were a single, neat, solvent with the weighted mean physical properties.

Apart from the use of technical grade solvents that are mixtures, situations may arise in which mixed solvents are produced involuntarily because of the systems themselves. This is the case in *liquid–liquid distribution* (solvent extraction), where two fairly immiscible liquids are in contact and a solute is allowed to distribute between the resulting two liquid phases. It is generally preferred to employ two solvents that are as immiscible with each other as possible, and one of the solvents is commonly water (or, rather, an aqueous solution of the solute or solutes to be extracted by the other solvent). Many useful systems of this kind involve a polar organic solvent that dissolves a considerable amount of water, forming a liquid phase that is still immiscible with the aqueous phase. Such a case is, for instance, the TBP mentioned before. When saturated by water at 25°C, the organic phase contains 49.7 mole% of water. In the case of *n*-butanol, however, the water-saturated organic phase contains 51.5 mole% of water, i.e., more water than butanol! This solvent is employed, for instance, in the removal of hydrochloric acid from aqueous mixtures of phosphoric acid and their calcium salts in manufacturing phosphoric acid. When water saturated, *n*-octanol, employed widely as a representative of biological membranes, contains 27.5 mole% of water. Values of the logarithm of the distribution ratio of solutes between *n*-octanol and water, used to express their hydrophilic (if negative) or hydrophobic (if positive) properties, normally pertain to this water-saturated organic phase.

On the whole, however, solvent mixtures are employed deliberately in order to gain the advantages that the two or more components confer on the mixture. In the laboratory practice of *liquid chromatography*, solvent mixtures are widely employed to permit optimal separation facilities for solutes distributing between the stationary and mobile phases, whether in normal or reversed-phase chromatography. There are several reasons for using solvent mixtures rather than neat solvents in liquid chromatography [especially high-performance liquid chromatography (HPLC)]. The most important one is the versatility in affecting selectivities and effecting separations when mobile-phase solvent mixtures are tailor made for the particular separation problem at hand. The composition of the solvent mixture employed is a degree of freedom added to the choice of the diverse solvents. The various solutes to be separated "perceive" differently the components of the mixed solvent, hence distribute differently between the mobile and stationary phases. Another reason is that changes in the composition of the solvent mixture can be made readily, applied stepwise or continuously (gradient elution). This can decrease greatly the total time needed for a separation without detrimentally affecting the resolution between the several solutes eluted from the column. An implicit reason, important in reversed-phase liquid chromatography (RPLC), is that one component (generally the less polar one, i.e., the organic component in aqueous mixtures) is selectively adsorbed in the stationary phase, modifying it favorably for the needed separation. The stationary phase in RPLC usually involves alkyl chains of silanoyl moieties and the mobile phase is usually an aqueous cosolvent mixture, hence the modification of the stationary phase by the cosolvent may play the key role in effecting a useful chromatographic separation. Finally, the modification of the physical characteristics of the mobile phase, such as its viscosity, may favor the use of an appropriately blended solvent mixture over the use of neat solvents.

Many other applications of solvents require the used of solvent mixtures rather than neat solvents, and in the following only a few are listed. *Solubilization* of certain solutes may be effected by the use of mixed solvents, not only because of their components solvating different parts of the solute molecules as mentioned before. One component may be chosen to solvate the solute and bring it into solution, and another may be used to convey desirable physical properties to the solution. Such a property may be a sufficiently high permittivity to permit electrolytic conductance and the application of electrochemical methods to the solute. For example, mixed solvents are commonly employed in high-energy-density batteries that are to function over a wide temperature range, including quite low temperatures. Thus, a mixture of 25 wt% propylene carbonate and 75 wt%

dimethoxyethane can be employed over the range -45 to $+25°C$ (or higher). The viscosity at the lowest temperature is 2.3 mPa·sec, the relative permittivity is 27, and these properties confer on lithium perchlorate solutions a maximal specific conductivity of 3.4 mS cm^{-1}, adequate for use in the battery. At the higher temperature (25°C), these values are 0.7 mPa·sec, 19, and 14.2 mS cm^{-1}, respectively, again in the desirable range of properties.

The opposite of solubilization is widely employed in the purification of the products of synthesis by *recrystallization*. The product may be formed in a solvent in which both it and the by-products and impurities are well soluble. It can then be precipitated out of solution by the addition of a certain other solvent, sometimes called an *antisolvent*. The by-products and impurities should be soluble in the resulting solvent mixture, whereas the desired product should be only sparingly soluble in it. This procedure may be repeated as often as required, and different cosolvents may be used, chosen for the purpose of removing specified impurities. Sometimes, such a process can be reversed, with the added solvent precipitating out undesirable by-products and leaving the desired product in solution. For instance, electrolytic by-products can be precipitated out when the added solvent drastically reduces the permittivity of the solution.

1.2 THE CONCEPT OF PREFERENTIAL SOLVATION

The term *solvation* was used in the introduction in its more colloquial sense, without being properly defined. On statistical thermodynamic grounds, solvation of a solute is defined as the process of transfer of a particle (atom, ion, aggregate, or molecule) of the moiety being solvated from a fixed position in an ideal gas phase to a fixed position in the liquid solvent solvating it [3]. This solvent may even be chemically identical with the solute itself, if liquid, the process then being condensation from the vapor. Generally, the solvent is some neat liquid, a mixture of liquids, or a solution already containing some of this solute alone or with other substances, with no restrictions on the concentrations. The fixed positions are specified in this definition in order to confine the process of solvation to the interactions of the solvated particle with its surroundings. The process is thus freed from the translational degrees of freedom of the solute, hence of the different volumes that the solute may occupy in the ideal gas phase and in the liquid solution. This convention takes care of the different standard states and concentration scales of the solute in the gas and in the liquid solution or mixture. With the process so defined, the thermodynamics of the solvation process pertain entirely to

1. The interactions of the solute particle with its environment in the liquid phase (because in the ideal gas phase it is devoid of interactions)
2. Changes in its internal degrees of freedom induced by these interactions
3. The effects of the solute particle on its environment due to its presence and interactions

As a first approximation only item 1 may be considered, but items 2 and 3 should not be ignored in a more rigorous treatment. The properties of the solvated solute may be quite different from those of the bare, non-solvated solute, as encountered in the ideal gas phase. Some spectro-scopic properties are, however, less affected than other properties by the solvation, hence by the nature and composition of the environment of the dissolved solute. For instance, the vibration frequencies of polyatomic solutes may depend rather weakly on the environment, if the interactions in item 1 are not very strong. However, even slight dependences can be used in order to shed light on such interactions due to solvation.

As a thought process, the solvation of atoms or molecules in a solvent or solvent mixture may be considered as proceeding in several virtual steps:

1. A cavity of appropriate size and shape is created in the solvent to accommodate the solute particle.
2. The solute particle is inserted into the cavity, without inter-acting with its surroundings.
3. The configuration of the solute molecule relaxes to its equili-brium state in the solution.
4. The solute particle interacts with its environment by disper-sion and dipole–induced dipole or dipole–dipole (quadrupole, etc.) interactions.
5. The molecules of the solvent or solvent mixture adjust them-selves to the presence of the solute in their midst.

All these steps, of course, take place simultaneously, but the ther-modynamic consequences may be thought of as being separate and addi-tive. Care must be taken not to reckon an effect more than once, if resulting from more than one step. If the molecules of the solute are rigid, or if it is atomic, then item 3 is inoperative. If the solute particles are flexible, then this item becomes important. In fact, with large polymeric solutes this can be a major item in the thermodynamics of the solvation. Conformational changes of such solutes, in particular the folding or unfolding of protein chains, are of crucial importance in practice. In general, the interactions listed in item 4 and the effects of the solute particle on its environment in item 5 are of a short range, proceeding over a few molecular diameters only. It should be remembered that the energies of dispersion forces decrease with r^{-6}, as do those of dipole–dipole interactions (when thermally averaged over all orientations), where r is the distance vector, in order to appreciate the short range of the effects.

Ionic solutes add the complication that charged particles cannot practically undergo the process of solvation just defined unless equivalent amounts of positive and negative charges are solvated simultaneously. Still, as a though process, a single ion of charge z_i (taken algebraically) may become solvated as described. The interactions and effects 4 and 5 are then of a considerably longer range, in particular if the "solvent" is already a solution containing ions. Ion-ion and ion-dipole interaction energies depend on r^{-1} and r^{-2}, respectively, the latter when not averaged over all dipole orientations. The coulombic interactions are not only of longer range but generally also stronger than those arising from dispersion forces alone. Over long distances away from an ion, the solvent may be approximated as a continuum characterized by its relative permittivity, but at short distances from it the molecular nature of the solvent comes into play.

All this was said from the viewpoint of the solute particle being solvated. However, steps 1, 4 and 5 in the virtual solvation process just described, involve the solvent, and if this is a mixed solvent there are important consequences to be considered. The foremost of these is whether or not the environment of the solute particle has the same composition as the bulk solvent mixture. In the latter case, there is an excess of the one solvent, hence a deficiency of the other in a binary solvent mixture in the solvation shells relative to the bulk composition. The latter case is designated as one of *preferential solvation*, or, if the preference proceeds to the extremity that one component is practically excluded from the vicinity of the solute, it is called *selective solvation*.

The molecules of the solvents in a homogeneous liquid mixed solvent interact with each other. These interactions may be stronger between different kinds of molecules than between those of the same kinds, and mutual attraction between the dissimilar solvent molecules takes place.

This is generally accompanied by a negative enthalpy of mixing; i.e., the mixing takes place exothermically with evolution of heat. Otherwise, self-interactions between molecules of the same kind predominate over those with alien molecules. If this preference does not go to extremes, the mixture remains homogeneous and does not split into two liquid phases. The preponderance of self-interactions in the mixture is generally manifested as a positive enthalpy of mixing, i.e., an endothermic one. In either case, these interactions may be weak, due to dispersion forces only, or may be stronger, if dipole–induced dipole, dipole–dipole, donor–acceptor, or hydrogen-bonding interactions take place between solvent molecules. Thus, the polarity, polarizability, electron pair donicity and acceptance, and hydrogen bond donicity and acceptance of the solvents come into play.

Consider now the case where the work of cavity formation in the mixed solvent, step 1 in the virtual solvation process, predominates over the Gibbs energies of all the other steps. If one solvent component has only weak dispersion interactions among its molecules, the work of cavity formation in it is minimal. It will then accumulate near the solute in order to minimize the overall Gibbs energy of the solvation process. Such a situation may be encountered in a solution of a solute with voluminous molecules, e.g., macromolecules of low polarity, in a mixture of an aliphatic hydrocarbon with a polar solvent. In such cases step 1 is automatically accompanied by step 5, and the solvent molecules redistribute themselves around the solute particle, which is then preferentially solvated by the hydrocarbon. Note that in such a case the preferential solvation is *not* due to stronger direct interactions of the solute with the hydrocarbon (its interactions with the polar component may be somewhat stronger). It arises because of the predominance of the cavity formation term for a very bulky solute that does not interact strongly with either solvent.

Generally, however, for solutes with small or moderately bulky molecules, the direct interactions of the solute with the solvent molecules in the solvent mixture, step 4, are the more important causes of preferential solvation. Charged solute particles interact preferentially with polar solvent molecules, the more so the higher the dipole moment of the latter. Such solutes may also interact preferentially with highly polarizable but nonpolar solvent molecules in which dipoles can be readily induced. Solute particles that can donate or accept a hydrogen bond interact strongly with solvent molecules that can, respectively, accept or donate them. They do so in particular with amphiprotic solvents, such as water, alkanols, or non- or singly substituted amides. Protic solute particles interact preferentially with polar aprotic solvents that are strong Lewis bases, i.e., have large electron pair donicities. It must be remembered that such solute–solvent interactions compete with solvent–solvent interactions, and a solute particle must overcome the latter if it is to be preferentially solvated in a solvent mixture. Thus, step 5 in the virtual solvation process is an essential part of the whole and must not be forgotten if the entire process is to be understood.

Consider now a solute that is identical to one component of a solvent mixture. In the simplest case, a particle of a component of a binary solvent mixture is the one that undergoes solvation in the mixture. All the previous considerations are still valid, and one may, therefore, speak of *preferential solvation in a binary solvent mixture*. The new particle will require a cavity to fit into, accommodate its configuration to having neighboring solvent molecules, organize around itself a solvation shell with which it interacts, and affect interactions between the solvent molecules in its surroundings, whether these are the same as itself or of the other kind. Whether preferential solvation will take place and whether self- or mutual interactions between solvent molecules prevail and, if so, to what extent depend on the natures of both solvents and their relative amounts.

The molecules of the two solvents in the mixture may be sufficiently alike in their chemical nature that the self- and mutual interactions balance each other well, their random mixing leads to the minimal Gibbs energy, and no preferential solvation takes place. However, if self-interactions prevail over mutual ones, the near solvation shell(s) of a solvent molecule will be constituted mainly by molecules of the same kind. Such a situation leads to clustering of at least one of the components and possibly of both, leading to *microheterogeneity*. If only one component is apt to aggregate to clusters, individual molecules of the other will be adjacent to the clusters. As the distance from a given molecule increases and concentric shells in the solvent mixture much farther from a molecule in a cluster are considered, their composition approaches that of the bulk solvent with its normal fluctuations. If both solvents aggregate, then under certain conditions phase separation may take place. On the contrary, if the mutual interactions prevail over the self-interactions, then the nearest solvation shell of a given solvent molecule may have an excess of the other kind. In certain cases, adduct formation to 1:1 or other simple stoichiometries may constitute this preference. This is especially the case when (electron pair) donor-acceptor and hydrogen bonding are the dominant interactions.

Preferential solvation in binary solvent mixtures manifests itself in many ways. The excess and deficiencies of molecules in certain regions of space near a given molecule can be discerned by structure determination methods, such as X-ray or neutron diffraction, or may be ascertained by computer simulations or by other methods described in this book. The interactions leading to the preferential solvation are manifested as thermodynamic excess functions: excess Gibbs energy of mixing, its first derivatives with respect to pressure (excess volume) and temperature (excess entropy and enthalpy), its second derivatives (excess compressibility, heat capacity, and expansivity), etc. In this connection the "excess" may be positive or negative. These quantities are measurable and permit conclusions to be drawn, indirectly, concerning the preferential solvation. Other measurable quantities, such as the relative permittivity and transport properties such as viscosity, as well as spectroscopic quantities such as vibrational or NMR spectra, are also relevant in this connection.

The present book discusses in detail these experimental approaches to ascertaining whether preferential solvation takes place and, if so, to what extent. Also, the consequences of preferential solvation, detected by a given experimental approach, for other properties of systems involving mixed solvents are described. Binary solvent mixtures without extraneous solutes are discussed first in Chapter 4, and ternary mixtures, consisting of a binary mixed solvent with an additional foreign solute, are discussed in Chapter 5. Extension of these concepts to multicomponent liquid system are more briefly discussed in Chapter 6.

1.3 THE COMPOSITION OF MIXED SOLVENTS

Solvent mixtures are sometimes prepared on a weight basis and sometimes on a volume basis. Because for many purposes the composition on the basis of amounts of substance (moles) is required, these compositions and their interrelationships should be well specified. The most straightforward is the composition on the weight basis, where weighed amounts W_i of the components ($i = A, B, C, \ldots$) are mixed. Such a composition is, of course, practically identical to that on a mass basis. For binary mixtures of solvents A and B only, just the weight (mass) percentage of one component needs to be specified: wt% $A = 100\, w_A$, where w_A is the *mass fraction* of solvent A. The corresponding value for solvent B is then $w_B = 1 - w_A$. In general the mass fraction w_i of component i is

$$w_i = W_i/(W_A + W_B + \cdots) = W_i/\sum_j W_j \qquad (1.1)$$

where the index j varies over all the components, including i. Such compositions can be readily specified to a reproducibility and accuracy of 1 part in 10,000.

For most solvents in common use the formula weight is known, numerically identical to the molar mass, M, commonly expressed in g mol^{-1} (or in kg mol^{-1} if strict adherence to the SI specifications is required). The *mole fraction x_i*, of component i is obtained when n_j moles of all the components are mixed to give a total of n moles of the mixture, whence

$$x_i = n_i/(n_A + n_B + \cdots) = n_i/\sum_j n_j = n_i/n \tag{1.2}$$

However, becuase for each component the number of its moles is proportional to the weight of it taken, $n_j = W_j/M_j$, Eq. (1.2) can be rewritten as

$$x_i = W_i/\sum_j W_j(M_i/M_j) \tag{1.3}$$

For a binary mixture of solvents A and B this relationship can be restated as

$$x_A = w_A/[w_A + (1 - w_A)M_A/M_B] \tag{1.4}$$

Conversely, if the mole fractions are known and it is desired to convert them into mass fractions, then for a binary mixture of solvents A and B:

$$w_A = x_A/[x_A + (1 - x_A)M_B/M_A] \tag{1.5}$$

[note the reversal with respect to Eq. (1.4) of the molar masses in the denominator]. The mole fractions generally differ appreciably from the mass fractions, in particular for aqueous solvent mixtures, because the molar mass of water is lower than that of most cosolvents, in the majority of cases considerably so. On going from pure water up to, say, 90 wt%cosolvent with a molar mass of 90 g mol^{-1}, the mole fraction of the latter, 0.643, will be much lower than its mass fraction. Mole fractions, when ≥ 0.1, are generally accurate to better than 1 part in 1000 and are often stated to four significant digits. If a component of the solvent mixture is a technical grade solvent (e.g., kerosene) that is itself a mixture of nonisomeric components, then its molar mass has to be specified as the weighted average of those of its constituents, possibly to no better than $\pm 2\%$. The mole fractions in the mixture cannot then be known to 1 part in 1000.

Solvent mixtures are commonly and conveniently prepared in the laboratory by adding a certain volume of one solvent (with a pipette or from a burette) into a volumetric flask and making up to volume with the other solvent. Such a procedure is accompanied by loss of accuracy in the composition of the mixed solvent because solvents generally shrink or expand on mixing. Only if the temperature is carefully controlled and specified volumes of the components are mixed can the knowledge of the densities of the neat solvents at the given temperature permit the accurate specification of the composition. However, for many applications the change of volume on mixing can be ignored and binary mixtures of solvents A and B prepared on a volume basis are reported in the literature as vol% A $= 100\ \phi_A$, where ϕ_A is the *volume fraction*.

The molar volume of a neat solvent j is $V_j = M_j/d_j$, where d_j is its density. The volume fraction of component i in a mixture is therefore (if volume changes on mixing are ignored)

$$\phi_i = n_i V_i / \sum_j n_j V_j \tag{1.6}$$

In a binary mixture of solvents A and B, the volume fractions are related to the mole and mass fractions according to

$$\phi_A = x_A/[x_A + (1 - x_A)(V_B/V_A)] \tag{1.7}$$

$$= w_A/[w_A + (1 - w_A)(d_B/d_A)] \tag{1.8}$$

$$x_A = \phi_A/[\phi_A + (1 - \phi_A)(V_A/V_B)] \tag{1.9}$$

$$W_A = \phi_A/[\phi_A + (1 - \phi_A)(d_B/d_A)] \tag{1.10}$$

In view of the uncertainty introduced by the shrinkage or expansion of the mixture relative to the volumes of the components being mixed, the compositions specified as volume fractions are generally accurate to no better than 1 part in 200. The mole fractions of cosolvents in aqueous mixtures are generally considerably lower than their volume fractions because the molar volume of water is smaller than the molar volumes of the common cosolvents.

There are two further measures of the composition of mixed solvents, and these are applied mainly to solutions in them. One is the *molality* (symbol m), i.e., the number of moles of a component B, C, ... in a unit mass (1 kg) of a given component A, designated as the solvent. If A is water, then the molality corresponds to the number of moles of components B, C, ... per $1/(0.018015\ \text{kg mol}^{-1}) = 55.51$ moles of water. For a mixture of cosolvent or solute B with water $x_B = m_B/(55.51 + m_B)$. For completely miscible solvents, the molality of a component may tend to infinity if the component designated as "solvent" becomes very dilute. Hence, the molality is rarely applied to a binary solvent mixture, where there is no good reason to consider one or the other component as the solvent. However, this measure is useful and widely used for solutions of substances in a mixed solvent of constant composition that is considered to behave as a single, quasi-neat, solvent. When the solvent mixture is one of light water and heavy water, the concept of *aqua-molality* is sometimes employed. This is the number of moles of solute per 55.51 moles of water (irrespective of its isotopic composition). That is, the aqua-molality is the amount of substance that would be in the solution per 1 kg of solvent if all the heavy water, D_2O, in the solvent mixture were exchanged for an equivalent amount (on a mole basis) of H_2O.

The other widely employed measure of the composition in solvent mixtures is the *molarity* (symbol c), i.e., the number of moles of components A, B,... per unit volume of the mixture, taken as 1 dm^3 ($= 1$ L, rather than the proper SI unit of $1 \, m^3$). This measure is, again, rarely used for binary solvent mixtures but may be useful in pointing out the limit, i.e., the maximal achievable concentration, of a component. This quantity is the reciprocal of the molar volume at the temperature under consideration: $c_{i \, max} = 1000/(V_i/cm^3 \, mol^{-1})$. More useful, in the fields of statistical mechanics and statistical thermodynamics, is the measure of *number density* (symbol ρ), i.e., the number of particles of a component per unit volume. If the unit volume considered is 1 cm^3, then $\rho_i = (N_{Av}/1000)c_i$; that is, the number density is directly proportional to the molarity. When other substances dissolved in solvent mixtures are considered, the number density or molarity of the solutes in the solution is of great significance and the fact that the solvent is a mixture can sometimes be ignored.

The different measures of composition of a binary solvent mixture should be borne in mind. These differences are enhanced when the densities and molar masses of the components differ widely. As an (extreme) example, measures of the compositions of mixtures of ethanol (A) and bromoform (B) are shown in Figure 1.1. The molar masses are 46.07 and 252.73 g mol^{-1} and the densities are 0.7892 and 2.8909 g cm^{-3} at 20°C, so that the molar volumes are 58.37 and 87.42 cm^3 mol^{-1}, respectively. The molality of component B, m_B, diverges toward infinity as the mole fraction x_B tends toward unity. The maximal molarity c_B of bromoform is 12.75 mol dm^{-3}, the reciprocal of its molar volume. The ratio

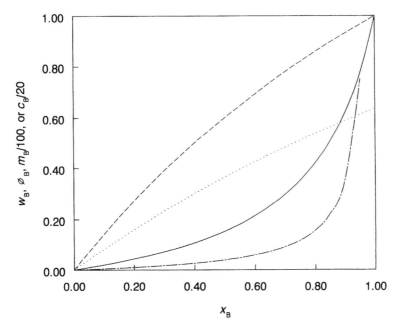

FIGURE 1.1 Various measures of the composition of a binary solvent mixture of ethanol (A) and bromoform (B) against the mole fraction x_B of the latter at 20°C: (———)w_B, (– – – – –) φ_B, (– · — · — · —·)$m_B/100$, and (············)$c_B/20$. (Note that the displayed volume fraction φ_B and the molarity c_B do not take volume changes on mixing into account.)

between the molar volumes is much smaller than that between the molar masses; therefore the φ_B curve is much closer to the diagonal than the w_B one.

REFERENCES

1. Y. Marcus. Introduction to Liquid State Chemistry. Chichester, UK: Wiley, 1977.
2. Y. Marcus. The Properties of Solvents. Chichester, UK: Wiley, 1998.
3. A. Ben-Naim. J Phys Chem 82:792, 1978; A. Ben-Naim, Y. Marcus. J Chem Phys 81:2016, 1984; A. Ben-Naim. J Solution Chem 30:475, 2001.

2

Properties of Binary Solvent Mixtures

2.1 REPRESENTATIVE SOLVENT MIXTURES

For the rational use of solvent mixtures it is essential to know their physical and chemical properties as functions of their composition and the ambiance. In view of the many hundreds of solvents in use, the binary combinations alone would come to tens of thousands, hence it is impossible to include a comprehensive survey of the published information in a book such as this. It is therefore necessary to limit the discussion to representatives of the major classes of solvents in common use. It is also expedient to present separately the data for aqueous mixtures, for solvents that have unlimited mutual solubility with water, and for mixtures of solvents not involving water.

In aqueous mixtures, the solvents that are completely miscible with water may be called its *cosolvents*. Generally, only the lower homologues of a series of aliphatic solvents are miscible with water, and very few aromatic solvents are. The complete miscibility of solvents with water depends on their hydrophilic character, expressible as the Leo and Hansch [1] log *P*. This is the logarithm of the partition constant (at infinite dilution) of the substance considered as a solute between

16

(water-saturated) 1-octanol and water. As a rule, the solvents with known negative values of log P or with negative values that can be assigned to them from group contributions are miscible with water. On the contrary, most of the solvents with log $P > 1.3$ have limited mutual solubility with water. Of those with in-between values, $0 \leq \log P \leq 1.3$, some are miscible (e.g., piperidine, log $P = 0.85$; dichloroacetic acid, log $P = 1.29$) and some are not (e.g., nitroethane, log $P = 0.18$; cyclohexanone, log $P = 0.81$). An outstanding exception to this rule is nitromethane, log $P = -0.34$, which is not completely miscible with water.

Concerning the alcohols, it is of interest to see (Sections 2.2 and 2.3) how the replacement of hydrogen atoms on the methyl group of methanol, CH_3OH, with other methyl groups to produce ethanol, CH_3CH_2OH, 2-propanol (*iso*-propanol), $(CH_3)_2CHOH$, and 2-methyl-2-propanol-(*tert*-butanol), $(CH_3)_3COH$, affects the properties of their aqueous mixtures. In the alcohols, furthermore, replacement of the methyl groups in ethanol and 2-propanol with $HOCH_2-$groups produces 1,2-ethanediol and glycerol, water-miscible solvents with some unique properties due to the proliferation of hydrogen-bonding sites.

Similarly, in the amides, the replacement of hydrogen atoms on the nitrogen atom of formamide, $HC(O)NH_2$, with methyl groups to produce N-methylformamide (NMF), $(HC(O)N(H)CH_3)$, and N,N-dimethylformamide (DMF), $HC(O)N(CH_3)_2$, has interesting effects. Among other amides, N-methylpyrrolidin-2-one and hexamethyl phosphoric triamide are commonly used as cosolvents. Only cyclic ethers, such as tetrahydrofuran (THF) and 1,4-dioxane(DX) among the common ones, are miscible with water, and only acetone (Me_2CO) among ketones is, but no common aliphatic esters are miscible with water. Common esters of acids other than carboxylic, such as carbonic, sulfuric, or phosphoric, are not miscible with water either. Formic and acetic acids represent the water-miscible carboxylic acids. Although in dilute solutions in water they are dissociated to ions to a large or small extent, for suitable solutes the aqueous mixtures can be used as solvents without their being protonated.

Of the aliphatic amines that may be used as solvents, triethylamine was selected, but it is miscible with water only below the (lower) consolute temperature of 18°C, but the aromatic amine pyridine is miscible with water and is included. Other nitrogen-containing solvent classes are nitriles, of which acetonitrile is a widely used cosolvent, but nitromethane, the lowest homologue of the nitro-compounds, has only limited mutual solubility with water, as noted before. Finally, sulfur compounds are represented by by dimethylsulfoxide, which is widely used as a cosolvent. In all, these 20 aqueous cosolvent mixtures should give a representative picture of the variation of the properties with the chemical nature. The properties of the aqueous mixtures are written throughout this book with water being component A and the cosolvent being component B, in terms of the mole fraction of B. The names of these solvents are abbreviated throughout this book as listed in the Preface.

The miscibility of nonaqueous solvents can be described in terms of the Hildebrand *solubility parameters* δ by the condition that $|\delta_A - \delta_B| < 4 \text{ MPa}^{1/2}$. Here $\delta_A = [(\Delta_V H_A - RT)/V_A]^{1/2}$, where $\Delta_V H_A$ is the enthalpy of vaporization of solvent A and V_A is its molar volume, and similarly for solvent B. The selection of solvent mixtures that do not involve water must also be severely limited but should include representatives of the common classes of solvents. It is not expedient to deal with mixtures of each of the 19 solvents selected for discussion here with all the others, i.e., 342 mixtures, mainly in view of the non availability of data. Only seven solvents are selected—benzene, tetrachloromethane, acetone, acetonitrile, nitromethane, methanol, and ethanol—as representatives of nonpolar, polar aprotic, and protic solvents. Methanol is the most water-like of these, hence is included beside the more organic-like ethanol. Their mixtures with the 18 other solvents from similar classes but with more diverse chemical functional groups are treated. These include in addition to the six from the preceding list *n*-hexane, *c*-hexane, toluene, chloroform, diethyl ether, tetrahydrofuran, 1,4-dioxane, ethyl acetate, pyridine, nitrobenzene, *N,N*-dimethylformamide, and dimethylsulfoxide. The list of solvents is, again, dictated mainly by the availability in the literature of data for the properties of interest. The names of these solvents are abbreviated throughout this book as listed in the Preface.

The most relevant physical and thermodynamic data for these mixtures are collected in tables dispersed throughout this chapter for the single temperature of 298.15 K (25°C), unless otherwise specified, and the ambient pressure of ≈ 0.1 MPa (1 atmosphere) as far as could be readily found. The reader may find data for temperatures other than 298.15 K (sometimes also for other pressures) in the cited references. The physical data for the aqueous solvent mixtures are summarized by the coefficients of suitable expressions, defined in the text. Those for the nonaqueous mixtures pertain to mixtures of each of the selected seven solvents A at definite mole fractions x_B: 0.00, 0.15, 0.35, 0.50, 0.65, 0.85, and 1.00 of the 18 solvents B (where known). These data permit the reconstruction of the curve of the given property against the composition, unless there are anomalies, which for some systems take place in the dilute

B (A-rich, $x_B < 0.15$) or B-rich ($x_B > 0.85$) regions. For these cases the original literature should be consulted. The thermodynamic data for both the aqueous and nonaqueous solvent mixtures are summarized by the coefficients of suitable expressions, defined in the text. The reader may find data for further solvent mixtures, not included in the tables in this book, in the cited references for solvents homologous with those that are included.

Only data for binary mixtures are presented as systematically as the scope and the preparation of this book permitted. Multicomponent solvent mixtures had to be excluded from detailed presentations of data due to their nonavailability in a systematic manner and to space limitations.

2.2 PHYSICAL PROPERTIES

Binary solvent mixtures exist as liquids, i.e., can be used as solvents, over a wide range of temperatures and pressures. The maximal liquid range of binary solvent mixtures at ambient and lower pressures extends from the lowest *eutectic* temperature and composition for the freezing mixture to the maximal *azeotropic* temperature and composition for the boiling mixture. It should be noted, however, that this range cannot cover the entire composition of the binary mixture; on the contrary, most of this temperature range covers only a narrow composition interval. Furthermore, not every mixture exhibits such singular points because some mixtures may freeze to solid solutions, not form an azeotrope, or form an azeotrope with a minimal boiling temperature. There are quite detailed data concerning the liquid ranges of nonaqueous solvent mixtures, compared with the surprising paucity of such data for aqueous mixtures.

Some solvent mixtures may form one or more crystalline compounds that fuse *congruently* to melts (liquid solvent mixtures) of the same composition. These have higher melting points than compositions away from the stoichiometric ones. Intersolvent crystalline compounds may, however, melt incongruently, that is, decompose on melting and deposit a solid of different composition. There may, again, exist more than one eutectic that is formed in such systems, having lower melting points than compositions away from the eutectic ones. Listed in Table 2.1 are the minimal eutectic temperatures and compositions.

The boiling point of many binary solvent mixtures is above that of the lower boiling component and over most of the range lower than that of the higher boiling one. Only over a small composition range rich in the

TABLE 2.1 Eutectic Temperatures, t_{eu}/°C, and Compositions of the Solvent in the Column Headings (wt% except where marked by m, where it is mole%) in Mixtures with the Solvent in the Left-Hand Column

Cosolvent	C_6H_6	CCl_4	MeOH	EtOH	Me_2CO
Hexane	−99.5, 10.7				
Cyclohexane	−41.9, 25.6	−30.8, 20.1			
Benzene		−40.9, 80 m	No eutectic	No eutectic	−96.0, 96.6
Toluene					
Chloroform	−80.7, 17.4	−81.7, 61.4	−111.8, 87.6 m		−115, 70
Tetrachloromethane	−40.9, 20 m				−110.3, 86.1
Methanol	No eutectic			−119, 90 m	−116.6, 49.7
Ethanol	No eutectic	−119, 10 m	−129.6, 16.9	−129.6, 83.1	−119.1, 20.5
2-Propanol					
2-Methyl-2-propanol					
Ethylene glycol					
Ethyl ether		−122.5, 12.5 m	−126.1, 31.5	−125.6, 60.8	−127.5, 20.7
Tetrahydrofuran		−24.7, 94.8 m*			
Dioxane		−110.3, 13.9	−116.6, 50.3	−119.1, 79.5	
Acetone	−96.0, 3.4				
Formic acid					
Acetic acid					
Ethyl acetate		−90, 16.5 m		−118.5, 11.4	
Pyridine	−57.0, 25.3				
Nitromethane					
Nitrobenzene	−23.2, 42.6	−35.0, 11.1			
Acetonitrile					
N-Methylformamide					
Dimethylformamide					
Dimethylsulfoxide					

Data from J Timmermans. The Physico-chemical Constants of Binary Systems in Concentrated Solutions. Vols 1 and 2. New York: Interscience, 1959 and *Landoldt-Börnstein. Vol II/3. Berlin: Springer, 1956.

higher boiling component may the boiling point of the mixture be higher than the latter. Such mixtures produce a vapor phase that differs in composition from the liquid phase in equilibrium with it. This permits the separation of the two components by fractional distillation. However, many mixtures form an azeotrope, i.e., a mixture with equal compositions in the two phases (liquid and vapor), boiling at a certain constant temperature. The azeotrope may have a boiling point lower than those of both components or one that is higher than the boiling points of both components. In both cases, when heat is applied to the mixture, the composition of the boiling liquid mixtures changes in the direction of the azeotropic composition, and when this is reached, no further separation of the components by fractional distillation can take place. The compositions and boiling temperatures of such systems are shown in Table 2.2.

Under elevated pressures, solvent mixtures may remain liquid on heating up to their critical points. These are discussed further later under the thermodynamic properties of the mixtures.

Most of the discussion here of physical properties of binary solvent mixtures has, however, to be limited to ambient conditions at which the data are generally available in the literature. These conditions are 25°C (298.15 K) and 1 bar (0.1 MPa), the latter pressure being nearly 1 atmosphere (0.101325 MPa). Some temperature- and pressure-derivative functions of the physical properties are, however, also briefly discussed in order to permit extension of the information to near-ambient conditions. For the relevant properties of the constituent neat solvents see Refs. 2–4.

The density d of a binary mixture of solvents A and B is readily measured, to 1 part in 10^4 if only ordinary care is taken and to 1 part in 10^6 if very careful temperature control and sophisticated instrumentation are used. The density of water [specified as mean ocean water (MOW) in terms of its isotopic composition] at 298.15 K and 0.1 MPa is 0.997045 g cm^{-3}. The densities of the cosolvents of water considered here range from 0.7228 for triethylamine to 1.2582 for glycerol. The densities of the organic solvents that are immiscible with water and are considered here have a larger range, from 0.6549 for n-hexane to 1.5841 for tetrachloromethane. However, the densities of mixtures of the various solvents considered here are *not* the composition-weighted averages of the densities of the components. The composition-weighted average does apply, but only approximately, to the molar volume of the mixture in terms of the molar volumes of the components. Because, however, the mixing process generally involves shrinkage or expansion, for high accuracy the densities of mixtures have to be measured directly.

TABLE 2.2 Azeotropic Temperatures, t_{az}/°C, and Compositions (wt%, except where marked by m, where it is in mole%, or by v, where it is in vol%) of Mixtures of the Solvent in the Column Heading with Those in the Left-Hand Column at Atmospheric Pressure, 0.101 Mpa

Cosolvent	H_2O	C_6H_6	CCl_4	MeOH	EtOH	Me_2CO	MeCN	$MeNO_2$
Hexane	61.6, ?	Nonazeotrope	Nonazeotrope	50, 26 v	58.7, 21.0	49.7, 53.5	56.8, 25 v	62.0, 21
Cyclohexane	69.0, 9	77.6, 51.9	Nonazeotrope	54, 38	64.9, 40	53.0, 67	62.2, 33 v	69.5, 26.5
Benzene	69.3, 8.8		Nonazeotrope	57.5, 39.1	76.9, 31.7	Nonazeotrope	73, 34	79.2, 14
Toluene	84.1, 13.5	Nonazeotrope	Nonazeotrope	63.8, 69	76.7, 68	Nonazeotrope	81.1, 78 v	96.5, 55
Chloroform	56.1, 2.8	Nonazeotrope	Nonazeotrope	53.5, 35m[a]	59.2, 83.9 m[a]	64.4, 21.5		Nonazeotrope
Tetrachloromethane	66, 4.1	Nonazeotrope		57.5, 39.1	64.9, 20	56.1, 88.5	65.1, 17	79.2, 86
Methanol	Nonazeotrope	57.5, 60.9	55.7, 79.4		Nonazeotrope	55.5, 88	63.5, 81	64.3, 12.2
Ethanol	78.17, 4.0	76.9, 68.3	64.9, 80	Nonazeotrope		Nonazeotrope	72.5, 44	76.1, 29.0
2-Propanol	80.3, 12.6	71.9, 66.7	68.7, 81.9	Nonazeotrope	Nonazeotrope	Nonazeotrope	74.5, 52	79.3, 27.6
2-Methyl-2-propanol	79.9, 11.8	72.6, 64.6	71.1, 83	Nonazeotrope	Nonazeotrope	Nonazeotrope		80.0, 21.2
Ethylene glycol	Nonazeotrope	Nonazeotrope		Nonazeotrope				Nonazeotrope
Ethyl ether	34.2, 1.3	Nonazeotrope	Nonazeotrope	Nonazeotrope	Nonazeotrope	Nonazeotrope	Nonazeotrope	
Tetrahydrofuran	63.8, 4.6			60.7, 31.0	?, 10	55.7, 96[c]	65.8, 8[c]	
Dioxane	87.8, 18	Nonazeotrope	Nonazeotrope	Nonazeotrope	Nonazeotrope	Nonazeotrope	Nonazeotrope	100.6, 56.5
Acetone	Nonazeotrope	Nonazeotrope	56.1, 11.5	55.5, 12	Nonazeotrope		Nonazeotrope	Nonazeotrope
Formic acid	107.2, 22.6	71.1, 69	66.7, 81.5			Nonazeotrope	Nonazeotrope	97.1, 54.5

Acetic acid	Nonazeotrope	80.1, 98	76.6, 97	Nonazeotrope	Nonazeotrope	Nonazeotrope	Nonazeotrope	101.2, 96
Ethyl acetate	70.4, 8.5	Nonazeotrope	74.8, 57	62.3, 44	71.8, 31.0	Nonazeotrope	74.8, 23	Nonazeotrope
Pyridine	94, 57	Nonazeotrope	Nonazeotrope	Nonazeotrope	Nonazeotrope	Nonazeotrope	Nonazeotrope	<100.5, >85
Nitromethane	83.6, 23.6	79.2, 86	71.3, 83	64.3, 87.8	76.1, 71.0	Nonazeotrope	Nonazeotrope	
Nitrobenzene	98.6, 88 v	Nonazeotrope	Nonazeotrope	Nonazeotrope	Nonazeotrope			Nonazeotrope
Acetonitrile	76.5, 16.3	73, 66	65.1, 83	63.5, 19	72.5, 56	Nonazeotrope		Nonazeotrope
N-Methylformamide	Nonazeotrope			Nonazeotrope[b]				
Dimethylformamide	Nonazeotrope	Nonazeotrope						
Dimethylsulfoxide	Nonazeotrope							

[a] Landoldt-Börnstein, Vol. IV/3. Berlin: Springer, 1975.
[b] B Kaczmarek, A Radecki. J Chem Eng Data 34:195, 1989.
[c] J Gmehling, R Böts. J Chem Eng Data 41:202, 1996, where data for other systems and pressures are to be found too.
Source: Data from LH Horsley. Azeotropic Data. Advances in Chemistry Series. Washington, DC: American Chemical Society, Vol 6, 1952 (most data), 35, 1962, and 116, 1973 (corrections and additions).

The mean molar mass of the mixture is readily known to 1 part in 10^4 (depending on the accuracy to which the composition is known):

$$M = x_A M_A + (1 - x_A)M_B \qquad (2.1)$$

Hence, the mean molar volume of the mixture is obtainable from its directly measured density d to better than 1 part in 5000 in the optimal case:

$$V = M/d \cong x_A V_A + (1 - x_A)V_B \qquad (2.2)$$

This accuracy deteriorates considerably if the approximate equality of the last part on the right-hand side is applied, the volume change on mixing being ignored. If this approximation is inverted, the approximate density of the mixture is obtained:

$$d \cong M/[x_A V_A + (1 - x_A)V_B] \qquad (2.3)$$

Because density measurements are relatively easy to carry out, it is always preferable to measure the density of the mixture directly rather than estimate it via Eq. (2.3).

Other physical properties that are of importance for solvent mixtures include transport properties such as the viscosity, η, electrical properties such as the relative permittivity, ε, and optical properties such as the refractive index, n_D, among others. Further properties are included in the Section 2.3.

The dynamic viscosity, η, of solvent mixtures can be measured directly, although it is usually the kinematic viscosity, $v = \eta/d$, that is measured. This quantity is given, e.g., by the rate of flow of the liquid through a capillary under controlled conditions, generally compared with the rate at which a reference liquid of known viscosity flows. The product of kinematic viscosity v, with the density, d, of the solvent mixture then provides the dynamic viscosity, η. The viscosities of liquids can generally be measured accurately enough to three significant digits, provided the temperature control is adequate. Values of the dynamic viscosity of the common aqueous mixtures at 298.15 K are expressed as a function of the mole fraction of the cosolvent as

$$\eta/\text{mPa·sec} = (1-x_B)\eta_A + x_B\eta_B + x_B(1-x_B)[\eta_0 + \eta_1(1-2x_B)$$
$$+ \eta_2(1-2x_B)^2 + \eta_3(1-2x_B)^3 + \cdots] \qquad (2.4)$$

where $\eta_A = 0.893$ mPa·sec is the viscosity of water at 298.15 K. The coefficients η_j are shown in Table 2.3. For equimolar mixtures expression (2.4) reduces to $\eta/\text{mPa·sec} = 0.446 + 0.5\eta_B + 0.25\eta_0$. This may

TABLE 2.3 Coefficients η_j of the Dynamic Viscosities, $\eta = 0.893(1-x) + \eta_{cosolvent}x + x(1-x)[\eta_0 + \eta_1(1-2x) + \eta_2(1-2x)^2 + \eta_3(1-2x)^3]$ mPa·sec of Aqueous Cosolvent Mixtures at 298.15 K, Where x is the Mole Fraction of the Cosolvent

Cosolvent	Ref.	$\eta_{cosolvent}$	η_0	η_1	η_2	η_3
Methanol	a	0.556	2.50	2.72	1.135	−0.176
Ethanol	a	1.087	3.92	5.03	4.60	0.968
2-Propanol	a	1.94	4.02	5.01	4.61	1.835
2-Methyl-2-propanol	a	4.55	9.21	10.13	5.06	3.04
1,2-Ethanediol	b	17.14	−6.41	−0.200	0.890	−1.251
Glycerol	l	945	−1616	1032	−653	387.7
Tetrahydrofuran	h	0.481	0.815	2.61	5.29	4.92
Dioxane	d	1.196	2.44	4.28	3.01	0.62
Acetone	e	0.301	0.614	2.89	3.80	0.310
Formic acid	m	1.522	0.451	−0.216	−0.020	−0.215
Acetic acid	j	1.13	5.51	−0.765	−0.697	1.555
Triethylamine	k	0.385	6.8	8.6	11.0	14.2
Pyridine	j	0.88	4.90	3.66	1.005	
Acetonitrile	h	0.361	−0.271	0.456	0.904	0.881
Formamide	i	3.35	−1.97	0.664	−0.203	−0.177
N-Methylformamide						
N,N-Dimethylformamide	g	0.801	4.13	8.48	4.12	−7.24
N-Methylpyrrolidin-2-one	f	1.663	12.47	13.72	1.39	−1.87
Hexamethyl phosphoric triamide	i	3.21	13.57	25.76	28.99	8.44
Dimethylsulfoxide	c	1.995	7.96	−10.38	−1.80	9.26

[a]S Westmeier. Chem Technol 28:480, 1976.
[b]F Corradini, A Marchetti, M Tagliazucchi, L Tassi, G Tosi. Aust J Chem 48:103, 1995.
[c]A Sacco, E Matteoli. J Solution Chem 26:527, 1997.
[d]MBR Paz, AH Buep, M Baron. An Asoc Quim Argent 76:69, 1988.
[e]K Noda, M Ohashi, K Ishida. J Chem Eng Data 27: 326, 1982.
[f]P Assarsson, FR Eirich. J Phys Chem 72:2710, 1968.
[g]G Chittleborough, C James, B Steel. J Solution Chem 17:1043, 1988.
[h]TM Aminabhavi, B Gopalakrishna. J Chem Eng Data 40:856, 1995.
[i]S Taniewska-Osinska, A Piekarska, A Kacperska. J Solution Chem 12:717, 1983; S Taniewska-Osinska, M Jozwiak. J Chem Soc Faraday Trans 1 85:2141, 1989.
[j]J Mazurkiewicz, P Tomasik. J Phys Org Chem. 3:493, 1990.
[k]AN Campbell, SY Lam. Can J Chem 51:4005, 1973; at 17°C, data read from small-scale figure.
[l]Y-M Chen, AJ Pearlstein. Ind Eng Chem Res 26:1670, 1987.
[m]M Afzal, H Ahmad, Z Ali Pak. J Sci Ind Res 33:321, 1990; extrapolated from 27.5°C to 25°C.

serve as a guideline concerning the direction in which the viscosity of such mixtures varies with the composition because there are cosolvents with lower and with higher viscosities than water. Some cosolvents that are highly hydrogen bonded have much higher viscosities than water (10- or 100-fold higher). These cosolvents include alkanediols, glycerol, and di- and triethanolamine, as well as other liquids when ambient temperatures are near their melting points (e.g., c-hexanol or 2,4-dimethylphenol).

For nonaqueous binary mixtures the general relationship

$$\ln \eta = x_A \ln \eta_A + x_B \ln \eta_B + x_A x_B \eta_{AB} \qquad (2.5)$$

is useful, where η_{AB} is a constant, generally independent of the composition but not of the temperature, that can be calculated from group contributions [5]. It is expedient here to record the viscosity values at 298.15 K for mixtures of each of the selected seven solvents A at definite mole fractions x_B: 0.00, 0.15, 0.35, 0.50, 0.65, 0.85, and 1.00 of the 18 solvents B (where known) in Table 2.4. Note that the value of η at a certain x_B

TABLE 2.4A Viscosity $\eta/\text{mPa·sec}$ at 298.15 K of Mixtures of Benzene ($\eta = 0.603\,\text{mPa·sec}$) with Cosolvents at the Stated Mole Fractions of the Cosolvent

Cosolvent	Ref.	0.15	0.35	0.50	0.65	0.85	1.00
Hexane							
Cyclohexane	a	0.604	0.605	0.626	0.669	0.773	0.892
Toluene	T	0.629	0.613	0.603	0.601	0.595	0.583
Chloroform	T	0.583	0.575	0.562	0.559	0.548	0.540
Tetrachloromethane	T	0.630	0.684	0.729	0.774	0.835	0.883
Methanol	T	0.562	0.564	0.565	0.568	0.572	0.576
Ethanol	T	0.561	0.611	0.672	0.759	0.913	1.124
Ethyl ether	T	0.549	0.417	0.335	0.292	0.272	0.230
Tetrahydrofuran							
Dioxane	b	0.621	0.711	0.787	0.869	0.991	1.090
Acetone	T	0.518	0.453	0.412	0.377	0.339	0.315
Ethyl acetate	T	0.549	0.476	0.443	0.414	0.383	0.366
Pyridine	T	0.657	0.706	0.746	0.791	0.849	0.878
Nitromethane	a	0.598	0.578	0.570	0.573	0.601	0.646
Nitrobenzene	T	0.690	0.848	1.002	1.195	1.524	1.834
Acetonitrile							
Dimethylformamide	c	0.612	0.631	0.653	0.684	0.744	0.805
Dimethylsulfoxide	d	0.869	1.201	1.424	1.622	1.845	1.995

Refs. can be found on pgs. 32 and 33.

TABLE 2.4B Viscosity $\eta/mPa \cdot sec$ at 298.15 K of Mixtures of Tetrachloromethane ($\eta = 0.901$ mPa·sec) with Cosolvents at the Stated Mole Fractions of the Cosolvent

Cosolvent	Ref.	0.15	0.35	0.50	0.65	0.85	1.00
Hexane							
Cyclohexane	T	0.875	0.858	0.857	0.861	0.878	0.903
Benzene	T	0.835	0.774	0.729	0.684	0.630	0.599
Toluene							
Chloroform	T	0.800	0.710	0.656	0.613	0.568	0.540
Methanol	T	0.862	0.854	0.840	0.800	0.686	0.552
Ethanol	T	0.881	0.917	0.968	1.021	1.072	1.081
Ethyl ether							
Tetrahydrofuran							
Dioxane	b	0.936	1.037	1.089	1.118	1.124	1.090
Acetone	T	0.782	0.647	0.563	0.493	0.403	0.329
Ethyl acetate	m	0.669	0.583	0.526	0.476	0.420	0.386
Pyridine							
Nitromethane	m	0.700	0.647	0.611	0.579	0.540	0.515
Nitrobenzene	e	0.999	1.129	1.251	1.392	1.612	1.801
Acetonitrile	f	0.789	0.658	0.570	0.490	0.398	0.340
Dimethylformamide	m	0.757	0.761	0.751	0.730	0.685	0.638
Dimethylsulfoxide	m	0.851	0.992	1.089	1.180	1.290	1.365

Refs. can be found on pgs. 32 and 33.

(say, 0.35) of solvent B in the table of mixtures involving solvent A is necessarily the same as the value of η at $x_A = 1 - x_B$ (i.e., 0.65) of solvent A in the table involving various mixtures of solvent B. Values for the neat solvents B can be read from the column $x_B = 1.00$, where slight variations are due to different sources of the data for given solvents A. Examples of the variation of the viscosity of binary mixtures are shown in Fig. 2.1 for water-miscible aqueous alkanols at 298.15 K.

Other transport properties of solvent mixtures that are of interest, such as the heat conductivity or the self- and mutual-diffusion coefficients, cannot be discussed here due to lack of space as well as widely available data.

The relative permittivity, ε, of solvent mixtures can also be measured directly and is generally reported to three or four significant digits. Its values for the common aqueous mixtures at 298.15 K may be expressed in a manner analogous to Eq. (2.4):

$$\varepsilon = (1-x_B)\varepsilon_A + x_B\varepsilon_B + x_B(1-x_B)[\varepsilon_0 + \varepsilon_1(1-2x_B) + \varepsilon_2(1-2x_B)^2] \quad (2.6)$$

TABLE 2.4C Viscosity $\eta/\text{mPa}\cdot\text{sec}$ at 298.15 K of Mixtures of Methanol ($\eta = 0.551\,\text{mPa}\cdot\text{sec}$) with Cosolvents at the Stated Mole Fractions of the Cosolvent

Cosolvent	Ref.	0.15	0.35	0.50	0.65	0.85	1.00
Hexane				Miscibility gap			
Cyclohexane				Miscibility gap			
Benzene	T	0.562	0.564	0.565	0.568	0.572	0.603
Toluene	g	0.564	0.577	0.580	0.581	0.578	0.554
Chloroform	T	0.609	0.652	0.641	0.620	0.583	0.552
Tetrachloromethane	T	0.686	0.080	0.840	0.854	0.862	0.903
Ethanol	T	0.624	0.721	0.797	0.882	0.995	1.097
Ethyl ether	T	0.452	0.355	0.306	0.271	0.237	0.226
Tetrahydrofuran							
Dioxane	h	0.580	0.662	0.749	0.858	1.043	1.214
Acetone	T	0.519	0.435	0.391	0.367	0.358	0.346
Ethyl acetate	i	0.486	0.453	0.444	0.437	0.428	0.424
Pyridine	j	0.610	0.686	0.738	0.786	0.843	0.882
Nitromethane	k	0.507	0.491	0.492	0.503	0.543	0.613
Nitrobenzene	l	0.719	0.937	1.113	1.309	1.596	1.788
Acetonitrile	n	0.464	0.389	0.352	0.331	0.328	0.341
Dimethylformamide	j	0.582	0.632	0.670	0.708	0.758	0.796
Dimethylsulfoxide	o	0.658	0.859	1.057	1.294	1.673	1.991

Refs. can be found on pgs. 32 and 33.

where $\varepsilon_A = 78.36$ is the relative permittivity of water at 298.15 K. The coefficients ε_j are shown in Table 2.5. Most water-miscible solvents have relative permittivities higher than 12 at 298.15 K, notable exceptions being the cyclic ethers, tetrahydrofuran (THF), and 1,4-dioxane (DX). The relative permittivity of a mixed solvent is *not* the weighted average of those of the components (i.e., the coefficients ε_j are nonzero). This is illustrated for aqueous methanol (MeOH), acetonitrile (MeCN) and *N,N*-dimethylformamide (DMF) in Fig. 2.2, notable curvature being seen for the dependence of ε on the mole fraction of the cosolvent. However, for nonaqueous mixtures the relative permittivity is, in fact, in many cases the harmonic mean of those of the components. That is, the reciprocal of ε is given by the weighted average of the reciprocals of ε_A and ε_B, but the weighting factors may be the mole fractions, the mass fractions, or more commonly the volume fractions. In the latter case, $\varepsilon \approx V/[x_A V_A/\varepsilon_A + x_B V_B/\varepsilon_B]$. It is expedient to record the relative

TABLE 2.4D   Viscosity $\eta/\text{mPa·sec}$ at 298.15 K of Mixtures of Ethanol ($\eta =$ 1.083 mPa·sec) with Cosolvents at the Stated Mole Fractions of the Cosolvent

Cosolvent	Ref.	0.15	0.35	0.50	0.65	0.85	1.00
Hexane	p	0.966	0.800	0.671	0.543	0.386	0.283
Cyclohexane	p	1.055	1.036	1.020	0.990	0.912	0.890
Benzene	T	0.913	0.759	0.672	0.611	0.561	0.603
Toluene (24°C)	T	1.042	0.876	0.771	0.666	0.598	0.592
Tetrachloromethane	T	1.072	1.021	0.968	0.917	0.881	0.903
Chloroform	T	0.985	0.827	0.762	0.737	0.706	0.600
Methanol	T	0.995	0.882	0.797	0.721	0.624	0.565
Ethyl ether	T	0.953	0.732	0.570	0.429	0.293	0.226
Tetrahydrofuran							
Dioxane	q	0.946	0.887	0.893	0.931	1.042	1.194
Acetone	T	0.858	0.641	0.514	0.435	0.366	0.329
Ethyl acetate	i	0.790	0.593	0.511	0.463	0.442	0.424
Pyridine	T	1.059	0.959	0.906	0.872	0.861	0.879
Nitromethane	k	0.854	0.690	0.623	0.580	0.564	0.613
Nitrobenzene	l	1.069	1.161	1.255	1.358	1.546	1.788
Acetonitrile	r	0.882	0.633	0.469	0.340	0.267	0.342
Dimethylformamide	s	0.933	0.829	0.804	0.804	0.809	0.813
Dimethylsulfoxide	o	1.015	1.094	1.219	1.384	1.680	1.991

Refs. can be found on pgs. 32 and 33.

permittivity values at 298.15 K for mixtures of each of the selected seven solvents A at definite mole fractions x_B: 0.00, 0.15, 0.35, 0.50, 0.65, 0.85, and 1.00 of the 18 solvents B (where known) in Table 2.6. Values for the neat solvents B can be read from the column $x_B = 1.00$, where slight variations are due to different sources of the data for given solvents A.

Rather than deal with the relative permittivity, the molar polarization of the solvent mixture, P, may be discussed:

$$P = (M/d)(\varepsilon - 1)/(\varepsilon + 2) \cong \phi_A P_A + \phi_B P_B$$
$$= (M/d)[\phi_A(\varepsilon_A - 1)/(\varepsilon_A + 2) + \phi_B(\varepsilon_B - 1)/(\varepsilon_B + 2)] \qquad (2.7)$$

This quantity is nearly additive in the volume fractions of the molar polarizations of the components, P_A and P_B. The relative permittivity of the mixture, ε, may then be back-calculated from those of the components from Eq. (2.7).

TABLE 2.4E   Viscosity $\eta/\text{mPa·sec}$ at 298.15 K of Mixtures of Acetone ($\eta = 0.303\,\text{mPa·sec}$) with Cosolvents at the Stated Mole Fractions of the Cosolvent

Cosolvent	Ref.	0.15	0.35	0.50	0.65	0.85	1.00
Hexane	p	0.298	0.293	0.289	0.285	0.280	0.277
Cyclohexane	p	0.311	0.323	0.332	0.392	0.415	0.875
Benzene	T	0.339	0.377	0.412	0.453	0.518	0.603
Toluene							
Tetrachloromethane	T	0.403	0.493	0.563	0.647	0.782	0.901
Chloroform	T	0.371	0.418	0.459	0.497	0.531	0.540
Methanol	T	0.358	0.367	0.391	0.435	0.519	0.551
Ethanol	T	0.366	0.435	0.451	0.641	0.858	1.083
Ethyl ether							
Tetrahydrofuran							
Dioxane							
Ethyl acetate							
Pyridine							
Nitromethane							
Nitrobenzene	T	0.450	0.644	0.816	1.030	1.415	1.813
Acetonitrile							
Dimethylformamide							
Dimethylsulfoxide	t	0.397	0.585	0.802	1.085	1.563	1.998

Refs. can be found on pgs. 32 and 33.

Other electrical properties of solvent mixtures, such as the conductivities, are not necessarily additive. In any case, few reliable data are available because the conductivities are strongly affected by the presence of impurities.

Values of the refractive index at the sodium D line, n_D, of liquids at ambient conditions range between 1.25 for perfluoroalkanes and 1.74 for diiodomethane, that is, not over a very large range. It should be noted, on the other hand, that with proper instrumentation and temperature control, refractive index values can be determined to ± 0.00001. However, the differences of data reported by various authors for a given liquid are considerably larger, so that the values shown here are given to only ± 0.0001. The values for common aqueous mixtures at 298.15 K are shown in Table 2.7 in terms of the coefficients n_j of Eq. (2.8):

$$n_D = 1.3325(1-x_B) + n_{DB}x_B + x_B(1-x_B)[n_0 + n_1(1-2x_B) + n_2(1-2x_B)^2$$
$$+ n_3(1-2x_B)^3] \tag{2.8}$$

TABLE 2.4F   Viscosity $\eta/\mathrm{mPa\cdot sec}$ at 298.15 K of Mixtures of Acetonitrile ($\eta = 0.340\,\mathrm{mPa\cdot sec}$) with Cosolvents at the Stated Mole Fractions of the Cosolvent

Cosolvent	Ref.	0.15	0.35	0.50	0.65	0.85	1.00
Hexane							
Cyclohexane							
Benzene							
Toluene	u	0.380	0.425	0.462	0.497	0.536	0.554
Chloroform							
Tetrachloromethane	f	0.398	0.491	0.570	0.658	0.789	0.909
Methanol	n	0.328	0.331	0.352	0.389	0.464	0.539
Ethanol	r	0.267	0.340	0.469	0.633	0.882	1.083
Ethyl ether							
Tetrahydrofuran							
Dioxane							
Acetone							
Ethyl acetate	v	0.399	0.459	0.484	0.489	0.465	0.426
Pyridine	w	0.425	0.535	0.618	0.699	0.805	0.884
Nitromethane	x	0.378	0.425	0.465	0.507	0.567	0.616
Nitrobenzene	n	0.468	0.687	0.890	1.127	1.495	1.810
Dimethylformamide	w	0.388	0.437	0.487	0.562	0.702	0.802
Dimethylsulfoxide	n	0.420	0.606	0.818	1.093	1.556	1.977

Refs. can be found on pgs. 32 and 33.

The variation of n_D between that of water and those of the cosolvents is rather small, not more than 13% for those listed in Table 2.8, but is by no means monotonous. Hence, care must be exercised if values of n_D are to be used for the determination of the composition, as is sometimes done. That is, for certain mixtures a given value of n_D corresponds to two different compositions. Also available are the values for the common nonaqueous solvent mixtures, although only for very few mixtures involving acetonitrile or nitromethane. These values, where known at 298.15 K, are shown in Table 2.8 for mixtures of each of the solvents A with the 18 solvents B at the definite mole fractions x_B: 0.00, 0.15, 0.35, 0.50, 0.65, 0.85, and 1.00.

Approximate values of the refractive index of binary solvent mixtures that are not shown in these tables can be obtained from the presumed independence of the molar refraction of a liquid, $R_D \equiv V(n_D^2 - 1)/(n_D^2 + 2)$, form the manner in which the constituent atoms or groups are

TABLE 2.4G Viscosity η/mPa·sec at 298.15 K of Mixtures of Nitromethane ($\eta =$ 0.623 mPa·sec) with Cosolvents at the Stated Mole Fractions of the Cosolvent

Cosolvent	Ref.	0.15	0.35	0.50	0.65	0.85	1.00
Hexane				Miscibility gap			
Cyclohexane				Miscibility gap			
Benzene	a	0.599	0.574	0.571	0.577	0.598	0.616
Toluene							
Tetrachloromethane	a	0.654	0.702	0.748	0.791	0.862	0.940
Chloroform							
Methanol	k	0.543	0.503	0.492	0.491	0.507	0.547
Ethanol	k	0.564	0.580	0.623	0.690	0.854	1.085
Ethyl ether							
Tetrahydrofuran							
Dioxane							
Acetone							
Ethyl acetate	y	0.588	0.548	0.519	0.491	0.456	0.433
Pyridine							
Nitrobenzene							
Acetonitrile	x	0.567	0.507	0.465	0.425	0.378	0.345
Dimethylformamide							
Dimethylsulfoxide	m	0.588	0.706	0.820	0.956	1.171	1.358

[a] TM Aminabhavi, VA Aminabhavi, SS Joshi, RH Balundgi. Indian J Technol 29:545, 1991.
[b] SL Oswal, RP Phalak. Int J Thermophys 13:251, 1992; at 303.15 K.
[c] PC Gupta, M Singh. Indian J Chem. 40A:293, 2001.
[d] W Dechert, R Elsebrock, IK Hakim, M Stockhausen. Z Naturforsch A52:807, 1997.
[e] SS Joshi, TM Aminabhavi, RH Balundgi. Indian J Technol 29:541, 1991.
[f] MG Prolongo, RM Masegosa, I Hernandez-Fuentes, A Horta. J Phys Chem 88:2163, 1984.
[g] RK Wanchoo, J Narayan. Phys Chem Liq 25:15, 1992.
[h] P Garcia, M Postigo. An Asoc Quim Argent 85:209, 1997.
[i] PS Nikam, TR Mahale, M Hasan. J Chem Eng Data 41:1055, 1996.
[j] MS Bakshi, G Kaur. J Chem Eng Data 42:298, 1997.
[k] C-H Tu, S-L Lee, IH Peng. J Chem Eng Data 46:151, 2001.
[l] PS Nikam, MG Jadhav, M Hasan. J Chem Eng Data. 40:931, 1995.
[m] LS Manjeshwar, TM Aminabhavi. J Chem Eng Data. 32:409, 1987; at 348.15 K [where $\eta(CCl_4) =$ 0.744 mPa·sec].
[n] MS Bakshi, J Singh, H Kaur, ST Ahmad, G Kaur. J Chem Eng Data 41:1459, 1996.
[o] PS Nikam, MC Jadhav, M Hasan. J Chem Eng Data 41:1028, 1996.
[p] I-C Wei, RL Rowley. J Chem Eng Data 29:332, 1984.
[q] MSh Ramadan, AM Hafez, AA El-Zyadi. Phys Chem (Peshawar, Pakistan) 10:55, 1991.
[r] PS Nikam, LN Shirsat, M Hasan. J Chem Eng Data 43:732, 1998.
[s] A Ali, AK Nain, M Kamil. Thermochim Acta 274:209, 1996.
[t] CM Kinart, WJ Kinart. Phys Chem Liq 29:1, 1995.

TABLE 2.4   Footnotes Continued

[u] LH Blanco, EA Gonzalez. Phys Chem Liq 30:213, 1995.
[v] SL Oswal, NB Patel. J Chem Eng Data 40:840, 845, 1995.
[w] J Sing, MS Bakshi. J Chem Res (M) 1992: 1701.
[x] A D'Aprano, A Capalbi, M Iammarino, V Mauro, A Princi, B. Sesta. J Solution Chem 24:227, 1995.
[y] S-L Lee, C-H Tu. J Chem Eng Data 44:108, 1999.
[T] J Timmermans. The Physico-chemical Constants of Binary Systems in Concentrated Solutions, Vols. 1 and 2. New York: Interscience 1959.

connected and of their environment (state of aggregation) and the external conditions. Thus, if it is assumed that

$$R_D = (1 - x_B)R_{DA} + x_B R_{DB}$$
$$= (1 - x_B)V_A(n_{DA}^2 - 1)/(n_{DA}^2 + 2) + x_B V_B(n_{DB}^2 - 1)/(n_{DB}^2 + 2)$$

$$(2.9)$$

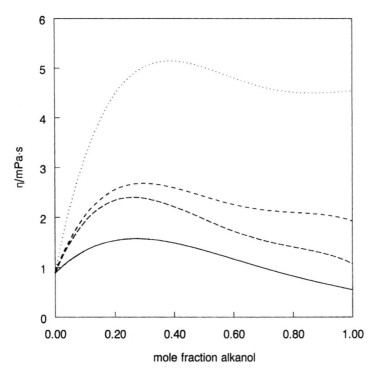

FIGURE 2.1   Viscosity at 298.15 K of aqueous MeOH (———), EtOH (— —), i-PrOH (-----), and t-BuOH (.........).

TABLE 2.5   Coefficients ε_j of the Relative Permittivities, $\varepsilon = 78.36\,(1-x) + \varepsilon_{cosolv}\,x + x(1-x)[\varepsilon_0 + \varepsilon_1(1-2x) + \varepsilon_2(1-2x)^2]$ of Aqueous Cosolvent Mixtures at 298.15 K, Where x is the Mole Fraction of the Cosolvent

Cosolvent	Ref.	$\varepsilon_{co\text{-}solv}$	ε_0	ε_1	ε_2
Methanol	a	32.66	−30.69	−1.80	
Ethanol	a	24.55	−58.65	−24.54	
2-Propanol	a	19.92	−89.84	−46.08	−52.62
2-Methyl-2-propanol	a	12.47	−111.62	−64.58	−111.41
1,2-Ethanediol	b	40.61	−28.01	−11.16	
Glycerol	a	42.50	−47.75	−20.28	
Tetrahydrofuran	c	7.58	−97.64	−50.92	−49.76
Dioxane	c	2.21	−121.54	−99.36	−74.64
Acetone	a	20.56	−71.96	−32.31	
Formic acid	j	57.20	33.93	7.50	
Acetic acid	j	6.60	−36.38	−12.21	
Triethylamine	l	2.42			
Pyridine	k	12.91	−98.23	−20.68	
Acetonitrile	d	35.94	−40.19	−13.21	
Formamide	e	109.50	53.65	29.17	28.06
N-Methylformamide	f	182.40	−51.71	16.11	
N,N-Dimethylformamide	g	36.71	−30.75	−0.67	
N-Methylpyrrolidin-2-one	h	32.20	−39.60	−9.87	
Hexamethyl phosphoric triamide	i	29.30	−102.68	−74.11	−65.05
Dimethylsulfoxide	f	46.45	2.05	7.06	

[a] G Akerlöf. J Am Chem Soc 54:4125, 1932.
[b] G Douheret, A Pal. J Chem Eng Data 33:40, 1988.
[c] G Akerlöf, OA Short. J Am Chem Soc 58:1241, 1936.
[d] C Moreau, G Douheret. J Chem Thermodyn 8:403, 1976.
[e] YM Kessler, VP Emelin, YS Tolube'ev, OV Truskov, RM Lapshin. Zh Strukt Khim 13:210, 1972.
[f] GG Karamyan, MI Shakhparonov. Zh Strukt Khim 22:54, 1981.
[g] G Douheret, M Morenas. Compt Rend C264:729, 1967.
[h] POI Virtanen. Suom Kemist B40:313, 1967.
[i] YM Kessler, VP Emelin, AI Mishustin, PS Yastremskii, ES Verstakov, NM Alpatova, MG Fomicheva, KV Kire'ev, VD Gruba, RK Bratishko. Zh Strukt Khim 16:797, 1975.
[j] U Kaaze, K Menzel, R Pottel. J Phys Chem 95:324, 1991.
[k] CK Hersh, GM Platz, RJ Swehla. J Phys Chem 63:1968, 1959.
[l] Y Marcus. The Properties of Solvents, Chichester,UK: Wiley, 1998; 293.15 K.

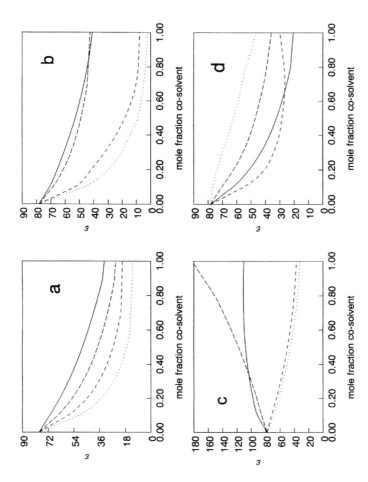

FIGURE 2.2 Relative permittivities of aqueous solvent mixtures at 298.15 K. Cosolvents: (a) (———) methanol, (———)ethanol, (----) 2-propanol, (.........)2-methyl-2-propanol; (b) (———) 1,2-ethanediol, (———) glycerol, (----)tetrahydrofuran, (.........) 1,4-dioxane; (c) (———) formamide, (----) N-methylformamide, (----) N,N-dimethylformamide, (.........) N-methylpyrrolidin-2-one; and (d) (———) acetone, (———) acetonitrile, (----) hexamethyl phosphoric triamide, (.........) dimethylsulfoxide.

TABLE 2.6A   Relative Permittivity ε at 298.15 K of Mixtures of Benzene ($\varepsilon = 2.282$) with Cosolvents at the Stated Mole Fractions of the Cosolvent

Cosolvent	Ref.	0.15	0.35	0.50	0.65	0.85	1.00
Hexane	T	2.214	2.111	2.054	2.010	1.950	1.904
Cyclohexane							
Toluene	T	2.311	2.321	2.337	2.355	2.365	2.378
Chloroform	T	2.549	3.002	3.268	3.625	4.099	4.654
Tetrachloromethane	T	2.271	2.262	2.253	2.244	2.237	2.230
Methanol	T	3.44	6.33	9.60	14.35	23.55	32.59
Ethanol	T	3.00	5.49	8.99	13.37	19.67	24.45
Ethyl ether	T	2.333	3.452	3.527	3.416	4.027	4.175
Tetrahydrofuran	a	2.923	3.691	4.300	5.023	6.302	7.600
Dioxane							
Acetone	T	3.75	6.21	8.66	11.74	16.38	20.87
Ethyl acetate							
Pyridine	T	3.36	5.07	6.52	8.12	10.47	12.41
Nitromethane	D	5.49	10.90	15.57	21.09	30.08	37.45
Nitrobenzene	T	5.84	11.11	15.36	20.37	28.36	34.12
Acetonitrile	D	5.14	10.05	14.44	19.77	28.79	36.80
Dimethylformamide							
Dimethylsulfoxide							

TABLE 2.6B   Relative Permittivity ε at 298.15 K of mixtures of Tetrachloromethane ($\varepsilon = 2.230$) with Cosolvents at the Stated Mole Fractions of the Cosolvent

Cosolvent	Ref.	0.15	0.35	0.50	0.65	0.85	1.00
Hexane							
Cyclohexane	b	2.181	2.135	2.103	2.072	2.034	2.007
Benzene	b	2.228	2.239	2.247	2.253	2.259	2.263
Toluene	b	2.245	2.278	2.302	2.323	2.348	2.364
Chloroform	T	2.497	2.881	3.215	3.605	4.226	4.770
Methanol	T	2.92	5.87	10.13	16.08	25.40	31.33
Ethanol	T	3.20	4.51	7.73	12.65	19.07	24.45
Ethyl ether							
Tetrahydrofuran							
Dioxane							
Acetone	T	3.83	6.53	9.13	12.03	16.54	20.87
Ethyl acetate	T	2.816	3.62	4.20	4.75	5.47	6.03
Pyridine							
Nitromethane	D	4.87	9.52	14.20	19.97	29.32	37.45
Nitrobenzene	T	5.71	11.48	16.52	21.92	30.01	36.50
Acetonitrile	D	4.33	8.49	12.67	18.05	27.65	36.80
Dimethylformamide							
Dimethylsulfoxide							

Refs. can be found on pg. 39.

TABLE 2.6C Relative Permittivity ε at 298.15 K of Mixtures of Methanol ($\varepsilon = 32.66$) with Cosolvents at the Stated Mole Fractions of the Cosolvent

Cosolvent	Ref.	0.15	0.35	0.50	0.65	0.85	1.00
Hexane				Miscibility gap			
Cyclohexane				Miscibility gap			
Benzene	T	23.55	14.35	9.60	6.33	3.44	2.29
Toluene							
Chloroform	D	26.53	18.64	13.68	9.84	6.61	4.80
Tetrachloromethane	T	25.40	16.08	10.13	5.87	2.92	2.23
Ethanol	D	31.82	29.82	28.583	27.33	26.00	24.55
Ethyl ether	D	24.33	15.86	11.59	8.82	6.58	4.35
Tetrahydrofuran							
Dioxane							
Acetone	D	30.47	27.12	25.26	23.89	22.42	20.87
Ethyl acetate	c	24.54	16.95	13.11	10.38	7.75	6.01
Pyridine	e	29.20	24.91	21.81	18.80	14.94	12.16
Nitromethane	D	33.46	33.58	34.07	34.76	35.93	37.45
Nitrobenzene	D	33.07	33.19	33.52	34.06	35.13	35.75
Acetonitrile	D	34.16	34.89	35.39	35.85	36.42	36.80
Dimethylformamide	e	35.58	37.71	38.28	38.28	37.88	37.61
Dimethylsulfoxide							

TABLE 2.6D Relative Permittivity ε at 298.15 K of Mixtures of Ethanol ($\varepsilon = 24.55$) with Cosolvents at the Stated Mole Fractions of the Cosolvent

Cosolvent	Ref.	0.15	0.35	0.50	0.65	0.85	1.00
Hexane	c	16.69	10.09	6.87	4.55	2.52	1.94
Cyclohexane							
Benzene	T	19.67	13.37	8.99	5.49	3.00	2.27
Toluene							
Chloroform	T	16.51	10.17	7.95	6.25	4.48	4.80
Tetrachloromethane	T	19.07	12.65	7.73	4.51	3.20	2.23
Methanol	D	26.00	27.33	28.58	29.82	31.82	32.66
Ethyl ether	T	18.86	13.78	10.47	17.76	5.57	4.29
Tetrahydrofuran							
Dioxane	d	19.11	11.12	7.41	5.24	3.03	2.24
Acetone	D	23.54	22.11	21.44	21.03	20.81	20.87
Ethyl acetate							
Pyridine							
Nitromethane							
Nitrobenzene	D	26.15	27.91	29.42	31.07	33.54	35.75
Acetonitrile							
Dimethylformamide	f	27.8	31.1	33.4	35.5	38.2	39.8
Dimethylsulfoxide							

Refs. can be found on pg. 39.

TABLE 2.6E Relative Permittivity ε at 298.15 K of Mixtures of Acetone ($\varepsilon = 20.87$) with Cosolvents at the Stated Mole Fractions of the Cosolvent

Cosolvent	Ref.	0.15	0.35	0.50	0.65	0.85	1.00
Hexane							
Cyclohexane							
Benzene	T	16.38	11.74	8.66	6.21	3.75	2.282
Toluene							
Tetrachloromethane	T	16.54	12.03	9.13	6.53	3.83	2.276
Chloroform	T	18.79	16.09	14.04	11.75	8.05	4.80
Methanol	D	22.42	23.89	25.26	27.12	30.47	32.66
Ethanol	D	20.81	21.03	21.44	22.11	23.54	24.55
Ethyl ether	D	17.68	13.74	11.20	8.89	6.07	4.35
Tetrahydrofuran							
Dioxane	D	17.00	12.56	9.68	7.05	3.98	2.22
Ethyl acetate	D	18.08	13.97	11.81	10.06	7.58	6.19
Pyridine							
Nitromethane	D	22.87	25.56	27.80	30.30	34.14	37.45
Nitrobenzene	T	23.60	26.89	29.06	31.05	33.51	35.22
Acetonitrile	D	23.13	25.76	27.91	30.27	33.78	36.80
Dimethylformamide	g	22.98	26.21	28.64	31.06	34.29	36.71
Dimethylsulfoxide	h	24.67	30.18	34.17	38.03	42.98	46.56

TABLE 2.6F Relative Permittivity ε at 298.15 K of Mixtures of Acetonitrile ($\varepsilon = 35.94$) with Cosolvents at the Stated Mole Fractions of the Cosolvent

Cosolvent	Ref.	0.15	0.35	0.50	0.65	0.85	1.00
Hexane							
Cyclohexane							
Benzene	D	28.79	19.77	14.44	10.05	5.14	2.82
Toluene	i	26.17	16.73	11.68	7.97	4.44	2.39
Tetrachloromethane	D	27.65	18.05	12.67	8.49	4.33	2.23
Chloroform							
Methanol	D	36.42	35.85	35.39	34.89	34.16	32.66
Ethanol	D	34.39	31.37	29.99	28.45	26.55	24.55
Ethyl ether	D	28.77	20.38	15.05	11.29	8.50	4.35
Tetrahydrofuran							
Dioxane	D	28.67	19.87	14.56	10.14	5.31	2.22
Acetone	D	33.78	30.27	27.91	25.76	23.13	20.87
Ethyl acetate	D	28.72	20.72	16.20	12.59	8.73	6.19
Pyridine	j	30.52	24.57	20.98	18.01	14.85	12.93
Nitromethane	k	36.01	36.11	36.21	36.31	36.40	36.48
Nitrobenzene	D	36.71	36.55	36.42	36.26	35.99	35.75
Dimethylformamide	j	35.45	35.20	35.27	35.55	36.15	36.73
Dimethylsulfoxide	l	37.65	39.81	41.41	43.00	45.10	46.66

TABLE 2.6G   Relative Permittivity ε at 298.15 K of Mixtures of Nitromethane ($\varepsilon = 35.87$) with Cosolvents at the Stated Mole Fractions of the Cosolvent

Cosolvent	Ref.	0.15	0.35	0.50	0.65	0.85	1.00
Hexane				Miscibility gap			
Cyclohexane				Miscibility gap			
Benzene	D	30.08	21.09	15.57	10.90	5.49	2.82
Toluene							
Tetrachloromethane	D	29.32	19.97	14.20	9.52	4.87	2.23
Chloroform							
Methanol	D	35.93	34.76	34.07	33.58	33.46	32.66
Ethanol	D	33.77	30.42	28.53	27.08	25.82	24.55
Ethyl ether	D	28.67	19.96	15.08	10.73	5.60	4.35
Tetrahydrofuran							
Dioxane							
Acetone	D	34.14	30.30	27.80	25.56	22.87	20.87
Ethyl acetate							
Pyridine							
Nitrobenzene	D	36.74	36.25	35.99	35.86	35.90	35.75
Acetonitrile	k	36.40	36.31	36.21	36.11	36.01	36.05
Dimethylformamide							
Dimethylsulfoxide							

[a]J Malecki, M Dutkiewicz. J Solution Chem 28:101, 1999.
[b]AH Buep. J Mol Liq 51:279, 1992.
[c]OB Rudakov. J Anal Chem 53:835, 1998.
[d]GE Papanastasiou, AD Papoutsis, GI Kokkinidis. J Chem Eng Data 32:3778, 1987.
[e]MS Baskhi, G Kaur. J Chem Eng Data 42:298, 1997.
[f]PW Khirade, A Chaudhari, JB Shinde, SN Helambe, SC Mehrotra. J Solution Chem 28:1031, 1999.
[g]L Kišovà, J Juřik, J Komenda. J Electroanal Chem 366:93, 1994.
[h]CM Kinart, WJ Kinart. Phys Chem Liq 29:1, 1995.
[i]LH Blanco, EA Gonzalez. Phys Chem Liq 30:213, 1995.
[j]J Sing, MS Bakshi. J Chem Res (M) 1992:1701.
[k]A D'Aprano, A Capalbi, M Iammarino, V Mauro, A Princi, B Sesta. J Solution Chem 24:227, 1995.
[l]MS Bakshi, J Singh, H Kaur, ST Ahmad, G Kaur. J Chem Eng Data 41:1459, 1996.
[D]D Decroocq. Bull Soc Chim Fr 1964:127.
[T]J Timmermans. The Physico-chemical Constants of Binary Systems in Concentrated Solutions, Vols 1 and 2. New York: Interscience, 1959.

TABLE 2.7 Refractive Indexes, n_D, of Aqueous Cosolvent mixtures at 298.15 K, at Specified Mole Fractions of the Cosolvent, x

Cosolvent	Ref.	0.15	0.35	0.50	0.65	0.85	1.00
Methanol	d	1.3392	1.3429	1.3422	1.3395	1.3343	1.3290
Ethanol	d	1.3536	1.3636	1.3654	1.3656	1.3644	1.3614
2-Propanol	b	1.3579	1.3729	1.3756	1.3746	1.3728	1.3751
2-Methyl-2-propanol							
1,2-Ethanediol	c	1.3699	1.3990	1.4108	1.4176	1.4239	1.4309
Glycerol	d	1.3939	1.4334	1.4482	1.4585	1.4695	1.4735
Tetrahydrofuran	a	1.3688	1.3928	1.3996	1.4011	1.4021	1.4071
Dioxane	b	1.3754	1.4027	1.4097	1.4109	1.4124	1.4201
Acetone	b	1.3534	1.3644	1.3649	1.3618	1.3569	1.3564
Formic acid							
Acetic acid	d	1.3578	1.3719	1.3765	1.3771	1.3748	1.3716
Triethylamine							
Pyridine							
Acetonitrile	a	1.3415	1.3459	1.3455	1.3436	1.3414	1.3420
Formamide							
N-Methylformamide							
N,N-Dimethylformamide	a	1.3783	1.4102	1.4200	1.4229	1.4241	1.4293
N-Methylpyrrolidin-2-one							
Hexamethyl phosphoric triamide							
Dimethylsulfoxide	b	1.3938	1.4389	1.4544	1.4608	1.4657	1.4745

[a] TM Aminabhavi, B Gopalakrishna. J Chem Eng Data 40:856, 1995.
[b] VI Lebed, FS Eddin. Zh Nevodnykh Rastvor 1:141, 1992.
[c] NG Tsierkezos, IE Molinou. J Chem Eng Data 43:989, 1998.
[d] DR Lide, ed. Handbook of Chemistry and Physics. 82nd ed. Baton Rouge, FL: CRC Press, 2001–2002, p 8–57 ff.

values of n_D of the mixture can then be obtained by back-calculation from the molar volumes (i.e., densities) and refractive indexes of the neat solvents A and B and the composition. Furthermore, since n_D is related to the square root of a linear function of R_D, any deviation from this assumed additivity rule of R_D is diminished in the derived n_D. The second equality on the right-hand side of Eq. (2.9) assumes a negligible excess volume of mixing and is responsible for a large measure of the observed deviations of the experimentally measured values of n_D from predictions of it by means of this expression. If the partial molar volumes v_A and v_B were used instead of V_A and V_B, better agreement would result. However, it ought to be easier to obtain accurate

TABLE 2.8A Refractive Index n_D at 298.15 K of Mixtures of Benzene ($n_D = 1.4979$) with Cosolvents at the Stated Mole Fractions of the Cosolvent

Cosolvent	Ref.	0.15	0.35	0.50	0.65	0.85	1.00
Hexane	T	1.4747	1.4453	1.4257	1.4087	1.3886	1.3750
Cyclohexane	a	1.4817	1.4624	1.4527	1.4425	1.4308	1.4236
Toluene	T	1.4969	1.4960	1.4954	1.4949	1.4943	1.4939
Chloroform							
Tetrachloromethane	T	1.4918	1.4837	1.4776	1.4715	1.4634	1.4572
Methanol	T	1.4850	1.4648	1.4450	1.4194	1.3729	1.3772
Ethanol	T	1.4825	1.4592	1.4397	1.4196	1.3871	1.3593
Ethylether	T	1.4764	1.4430	1.4233	1.4034	1.3670	1.3526
Tetrahydrofuran							
Dioxane							
Acetone	b	1.4790	1.4528	1.4320	1.4101	1.3795	1.3557
Ethylacetate	c	1.4808	1.4543	1.4349	1.4158	1.3908	1.3723
Pyridine							
Nitromethane							
Nitrobenzene	T	1.5087	1.5189	1.5266	1.5342	1.5445	1.5520
Acetonitrile							
Dimethylformamide							
Dimethylsulfoxide	d	1.459	1.463	1.473	1.476	1.478	1.479

TABLE 2.8B Refractive Index n_D at 298.15 K of Mixtures of Tetrachloromethane ($n_D = 1.4574$) with Cosolvents at the Stated Mole Fractions of the Cosolvent

Cosolvent	Ref.	0.15	0.35	0.50	0.65	0.85	1.00
Hexane							
Cyclohexane	c	1.4547	1.4477	1.4426	1.4375	1.4310	1.4262
Benzene	T	1.4634	1.4715	1.4776	1.4837	1.4918	1.4979
Toluene	e	1.4608	1.4693	1.4750	1.4803	1.4865	1.4905
Chloroform	T	1.4552	1.4532	1.4506	1.4480	1.4465	1.4439
Methanol	T	1.4516	1.4375	1.4211	1.3997	1.3635	1.3304
Ethanol	T	1.4475	1.4331	1.4202	1.4053	1.3814	1.3596
Ethyl ether	T	1.4420	1.4218	1.4061	1.3899	1.3677	1.3499
Tetrahydrofuran							
Dioxane							
Acetone	T	1.4459	1.4288	1.4142	1.3983	1.3753	1.3561
Ethyl acetate	c	1.4466	1.4288	1.4156	1.4025	1.3852	1.3723
Pyridine							
Nitromethane							
Nitrobenzene							
Acetonitrile							
Dimethylformamide							
Dimethylsulfoxide							

Refs. can be found on pg. 44.

TABLE 2.8C Refractive Index n_D at 298.15 K of Mixtures of Methanol ($n_D = 1.3265$) with Cosolvents at the Stated Mole Fractions of the Cosolvent

Cosolvent	Ref.	0.15	0.35	0.5	0.65	0.85	1.00
Hexane				Miscibility gap			
Cyclohexane				Miscibility gap			
Benzene	T	1.3729	1.4194	1.4450	1.4648	1.4850	1.4978
Toluene							
Chloroform	T	1.3592	1.3869	1.4046	1.4180	1.4310	1.4493
Tetrachloromethane	T	1.3635	1.3997	1.4211	1.4375	1.4516	1.4563
Ethanol	f	1.3331	1.3405	1.3456	1.3502	1.3556	1.3592
Ethyl ether	g	1.3354	1.3428	1.3460	1.3477	1.3488	1.3493
Tetrahydrofuran	h	1.3402	1.3519	1.3579	1.3625	1.3671	1.3698
Dioxane							
Acetone	T	1.3355	1.3438	1.3482	1.3512	1.3541	1.3557
Ethyl acetate	h	1.3474	1.3677	1.3793	1.3887	1.3988	1.4050
Pyridine	i	1.3780	1.4282	1.4573	1.4791	1.4969	1.5018
Nitromethane							
Nitrobenzene							
Acetonitrile	j	1.3292	1.3319	1.3337	1.3351	1.3369	1.3416
Dimethylformamide	i	1.3491	1.3729	1.3884	1.4018	1.4164	1.4250
Dimethylsulfoxide							

TABLE 2.8D Refractive Index n_D at 298.15 K of Mixtures of Ethanol ($n_D = 1.3594$) with Cosolvents at the Stated Mole Fractions of the Cosolvent

Cosolvent	Ref.	0.15	0.35	0.50	0.65	0.85	1.00
Hexane	k	1.3675	1.3654	1.3636	1.3618	1.3691	1.3568
Cyclohexane	h	1.3715	1.3864	1.3965	1.4056	1.4163	1.4235
Benzene	T	1.3871	1.4196	1.4397	1.4592	1.4825	1.4980
Toluene							
Chloroform							
Tetrachloromethane	T	1.3814	1.4053	1.4202	1.4331	1.4475	1.4577
Methanol	f	1.3556	1.3502	1.3456	1.3405	1.3331	1.3265
Ethyl ether	g	1.3587	1.3571	1.3554	1.3535	1.3508	1.3489
Tetrahydrofuran							
Dioxane	l	1.3721	1.3865	1.3957	1.4039	1.4136	1.4205
Acetone	m	1.3590	1.3585	1.3580	1.3574	1.3565	1.3558
Ethyl acetate							
Pyridine	T	1.3883	1.4251	1.4501	1.4686	1.4914	1.5068
Nitromethane							
Nitrobenzene							
Acetonitrile							
Dimethylformamide							
Dimethylsulfoxide							

Refs. can be found on pg. 44.

TABLE 2.8E   Refractive index n_D at 298.15 K of Mixtures of Acetone ($n_D = 1.3560$) with Cosolvents at the Stated Mole Fractions of the Cosolvent

Cosolvent	Ref.	0.15	0.35	0.50	0.65	0.85	1.00
Hexane	n	1.3692	1.3652	1.3626	1.3602	1.3575	1.3558
Cyclohexane	b	1.3660	1.3798	1.3900	1.4000	1.4132	1.4231
Benzene	b	1.3795	1.4101	1.4320	1.4528	1.4790	1.4979
Toluene							
Tetrachloromethane	T	1.3753	1.3983	1.4142	1.4288	1.4459	1.4572
Chloroform	T	1.3701	1.3888	1.4025	1.4153	1.4315	1.4430
Methanol	T	1.3541	1.3512	1.3482	1.3438	1.3355	1.3265
Ethanol	m	1.3565	1.3574	1.3580	1.3585	1.3590	1.3592
Ethyl ether							
Tetrahydrofuran							
Dioxane							
Ethyl acetate							
Pyridine							
Nitromethane							
Nitrobenzene							
Acetonitrile							
Dimethylformamide							
Dimethylsulfoxide							

TABLE 2.8F   Refractive Index n_D at 298.15 K of mixtures of Acetonitrile ($n_D = 1.3416$) with Cosolvents at the stated mole fractions of the Cosolvent

Cosolvent	Ref.	0.15	0.35	0.50	0.65	0.85	1.00
Hexane							
Cyclohexane							
Benzene							
Toluene							
Tetrachloromethane							
Chloroform							
Methanol	j	1.3369	1.3351	1.3337	1.3319	1.3292	1.3262
Ethanol							
Ethyl ether							
Tetrahydrofuran							
Dioxane							
Acetone							
Ethyl acetate							
Pyridine	o	1.358	1.385	1.407	1.432	1.471	1.506
Nitromethane	p	1.3475	1.3555	1.3613	1.3669	1.3741	1.3794
Nitrobenzene	j	1.3965	1.4530	1.4870	1.5138	1.5382	1.5482
Dimethylformamide	o	1.351	1.368	1.383	1.398	1.415	1.425
Dimethylsulfoxide	j	1.3679	1.3967	1.4166	1.4350	1.4571	1.4719

Refs. can be found on pg. 44.

TABLE 2.8G Refractive Index n_D at 298.15 K of Mixtures of Nitromethane ($n_D = 1.3796$) with Cosolvents at the Stated Mole Fractions of the Cosolvent

Cosolvent	Ref.	0.15	0.35	0.50	0.65	0.85	1.00
Hexane				Miscibility gap			
Cyclohexane				Miscibility gap			
Benzene							
Toluene							
Tetrachloromethane							
Chloroform							
Methanol							
Ethanol							
Ethyl ether							
Tetrahydrofuran							
Dioxane							
Acetone							
Ethyl acetate							
Pyridine							
Nitromethane							
Nitrobenzene							
Acetonitrile	p	1.3741	1.3669	1.3613	1.3555	1.3475	1.3413
Dimethylformamide							
Dimethylsulfoxide							

[a] M Iglesias, B Orge, J Tojo. J Chem Thermodyn 26:1179, 1994.
[b] A Tasić, BD Djordjević, K Grozdanić. J Chem Eng Data 37:310, 1992.
[c] TM Aminabhavi. J Chem Eng Data 32:406, 1987.
[d] W Dechert, R Elsebrock, M Stockhausen, Z Naturforsch A52:807, 1997.
[e] AH Buep. J Mol Liq 51:279, 1992.
[f] L Albuquerque, C Ventura, R Gonçalves. J Chem Eng Data 41:685, 1996.
[g] J Canosa, A Rodriguez, J Tojo. Fluid Phase Equil 156:57, 1999.
[h] OB Rudakov, VF Selemenev. Zh Fiz Khim 73:2019, 1999.
[i] MS Bakshi, G Kaur. J Chem Eng Data 42:298, 1997.
[j] MS Bakshi, J Singh, H Kaur, ST Ahmad, G Kaur. J Chem Eng Data 41:1459, 1996.
[k] B Orge, A Rodriguez, JM Canosa, G Marino, M Iglesias, J Tojo. J Chem Eng Data 44:1041, 1999.
[l] GE Papanastasiou, AD Papoutsis, GI Kokkinidis. J Chem Eng Data 32:377, 1987.
[m] M Iglesias, B Orge, M Dominguez, J Tojo. Phys Chem Liq 37:9, 1998.
[n] G Marino, MM Pineiro, M Iglesias, B Orge, J Tojo. J Chem Eng Data 46:728, 2001.
[o] J Sing, MS Bakshi. J Chem Res (M) 1992; 1701.
[p] A D'Aprano, A Capalbi, M Iammarino, V Mauro, A Princi, B Sesta. J Solution Chem 24:227, 1995.
[T] J Timmermans. The Physico-chemical Constants of Binary Systems in Concentrated Solutions, Vols 1 and 2. New York: Interscience, 1959.

experimental values of n_D than of the excess volume of mixing, so that if the approximation of Eq. (2.9) is not acceptable, then direct measurement should be resorted to.

2.3 THERMODYNAMIC PROPERTIES

Solvent mixtures behave ideally in rare cases and then are said to obey *Raoult's law*, according to which $p_A = x_A p_A^\circ$ and $p_B = x_B p_B^\circ$ over the entire composition range. The p_i° are the vapor pressures of the pure components. Near homologues, e.g., n-hexane+n-heptane, or positional isomers, e.g., 1,2-xylene+1,3-xylene, may approximate such behavior. Generally, however, solvents that are completely miscible with each other form rather nonideal mixtures (and the mixtures that show miscibility gaps are strongly nonideal). This nonideality is manifested by the *deviations from Raoult's law*, i.e., the partial vapor pressures p_i of the components are either not proportional to their mole fractions or, if proportional over certain composition ranges, the proportionality coefficients is not the vapor pressure of the pure component, p_i°. The deviations from Raoult's law for the binary mixtures of solvents A and B can be positive, $p_A > x_A p_A^\circ$, or negative, $p_A < x_A p_A^\circ$, as the case may be, but the deviations for solvent B are *not independent* of those of solvent A (see the following). When the deviations are positive the mixture may have a minimal boiling azeotrope (see Section 2.2), and when negative a maximal boiling azeotrope may result. In the former case, molecules of the two solvents prefer their own vicinity, and in the latter case, mutual attraction of the molecules of the two solvents takes place.

A rough guide to what may be expected from solvent pairs belonging to sets with characteristic functional groups is shown in Table 2.9, taken from Robbins [6]. Positive, negative, and very small deviations from Raoult's law are marked by $+$, $-$, and 0 in this table. The following classes of solvents are recognized:

Hydrogen bond donors: phenols (group 1); acids, thiols (group 2); alcohols, water (group 3); multihaloalkanes with active hydrogen atoms (group 4)

Hydrogen bond acceptors: ketones, disubstituted amides, sulfones, phosphine oxides (group 5); tertiary amines (group 6); secondary amines (group 7); primary amines, unsubstituted amides (group 8); ethers, sulfoxides (group 9); esters, aldehydes, carbonates, phosphates, nitrates, nitrites, nitriles (group 10); aromatic

TABLE 2.9 Deviations from Raoult's Law for Solvents Belonging to Various Classes Being Mixed with Other Solvents at Ambient Conditions (see text for the definition of the classes and the meanings of $+$, 0, and $-$) [6]

Class	1	2	3	4	5	6	7	8	9	10	11	12
1	0	0	−	0	−	−	−	−	−	−	−	−
2	0	0	−	0	−	−	0	0	0	0	−	−
3	−	−	0	+	0	−	−	−	−	−	−	−
4	0	0	+	0	−	−	−	−	−	−	0	−
5	−	−	0	−	+	0	−	−	0	+	0	0
6	−	0	−	−	+	+	0	0	0	0	0	−
7	−	0	−	−	+	+	0	0	0	0	0	−
8	−	0	−	−	+	+	0	0	−	+	−	−
9	−	0	+	−	+	0	0	−	0	+	0	−
10	−	0	+	−	+	+	0	−	−	0	−	−
11	+	+	+	0	+	0	0	−	0	+	0	0
12	+	+	+	+	+	0	+	+	+	+	0	0

hydrocarbons, olefins, haloaromatics, multihaloalkanes with no active hydrogen atoms (group 11)

Non−hydrogen-bonding solvents: paraffins, carbon disulfide (group 12)

Certain classes of solvents have not been considered in Table 2.9, such as nitro compounds, and bifunctional solvents are also excluded.

It is assumed here that the vapor pressures are sufficiently low for the vapor phase to be dilute and constitute an ideal mixture of the vapors. When this is the case, the *fugacities* (vapor activities, equal to the activities in the liquid mixture) are well approximated by the vapor pressures. Otherwise, nonideality of the vapor phase must be taken into account by means of the appropriate virial coefficients, that of the mixed vapor being well approximated by the composition-weighted mean of those of the components (see DIPPR [3] for values). If one component of the binary liquid mixture is practically nonvolatile at the given temperature, then the total measurable vapor pressure is that of the other component. If both components are volatile, then the vapor composition must be determined experimentally (e.g., by gas chromatography) in addition to the total vapor pressure $p = p_A + p_B$ in order to provide the partial vapor pressures. The vapor composition is generally designated in terms of the mole fractions y_i of the components in the vapor ($p_i = y_i p$) and is shown as curves $y_i(x_i)$. These are presented either isothermally, for varying total pressures of the vapor above the mixture, or isobarically, for varying temperatures, and in the latter case often for mixture boiling at atmospheric pressure.

In a binary mixture A + B, the partial vapor pressure of component A (or its fugacity, if a correction for vapor nonideality is required), p_A, is related to its *excess chemical potential* μ_A^E by

$$\mu_A^E = RT \ln (p_A/x_A p_A^\circ) = RT \ln f_A \qquad (2.10)$$

where $f_A = p_A/x_A p_A^\circ$ is its (rational, i.e., on the mole fraction basis) *activity coefficient*. Similar expressions pertain to component B. The *activity* of component A in the liquid mixture is $a_A = x_A f_A = p_A/p_A^\circ$ (equaling the relative vapor pressure) and is equal to its activity in the vapor phase due to the equilibrium between the two phases. The *Gibbs–Duhem relation* requires that at constant temperature and pressure in a binary mixture:

$$x_B \, d \ln \mu_B + x_A \, d \ln \mu_A = 0 \qquad (2.11)$$

hence also

$$x_B \, d \ln f_B + x_A \, d \ln f_A = 0 \qquad (2.12)$$

This puts a constraint on the variation of the activity coefficients (and partial vapor pressures) in the mixture. At the limits of the neat components the rational activity coefficients are unity, $\lim(x_A \rightarrow 1) f_A = 1$, and similarly for component B. As component B is added to neat component A, forming the mixture, f_A may decrease from unity but need not do so: it may become larger than unity. However, the change in f_A is necessarily accompanied by a corresponding change of f_B as prescribed by Eq. (2.12). At infinite dilution of a component, say A, in the other as the solvent, its activity coefficient has a definite value, f_A^∞, that can be readily measured by gas–liquid chromatography if component B is nonvolatile.

The nonideality of the entire liquid mixture, rather than that of its individual components, is described by its *molar excess Gibbs energy of mixing*, G^E. The excess chemical potentials are the partial derivatives of the excess Gibbs energy of mixing with regard to the composition. In a binary mixture:

$$x_A \, \mu_A^E + x_B \, \mu_B^E = G^E \qquad (2.13)$$

Vapor–liquid equilibrium (VLE) data are used to determine the excess Gibbs energy of mixing, which is thus obtainable from the total vapor pressure p and the vapor composition $y_A = 1 - y_B$ at varying compositions of the liquid mixture and given T. It is generally describable as the *Redlich–Kister polynomial* [7] in the composition:

$$G^{E} = x_{B}(1 - x_{B})[g_{0} + g_{1}(1 - 2x_{B}) + g_{2}(1 - 2x_{B})^{2} + g_{3}(1 - 2x_{B})^{3} + \cdots]$$

$$(2.14)$$

unless the $G^{E}(x_{B})$ curve is very skew. In that case the factor $[1 - g_{den}(1 - 2x_{B})]$ is placed in the denominator of Eq. (2.14), generally with a value of $g_{den} < 1$. This representation has the necessary property that $G^{E} = 0$ for both neat solvents, i.e., $x_{A} = 0$ ($x_{B} = 1$) and $x_{B} = 0$ ($x_{A} = 1$). The coefficients g_{0} to g_{3} for the aqueous mixtures of 20 common cosolvents at 298.15 K are shown in Table 2.10. For the nonaqueous mixtures the corresponding coefficients are given in Table 2.11. The coefficients with even indexes produce symmetrical curves of $G^{E}(x_{B})$, whereas those with odd indexes produce skewness in the curves. For the equimolar mixtures with $x_{B} = x_{A} = 0.5$ the excess molar Gibbs energy is $G^{E}(0.5) = g_{0}/4$.

The excess chemical potential of component A can now be written as a *k-suffix Margules expression*, a polynomial in the mole fraction of component B, where k is the maximal index of the coefficients in Eq. (2.14) plus two and is the highest exponent of the mole fraction in Eq. (2.15):

$$\mu_{A}^{E} = RT \ln f_{A} = g_{0}'x_{B}^{2} + g_{1}'x_{B}^{3} + g_{2}'x_{B}^{4} + g_{3}'x_{B}^{5} + \cdots \qquad (2.15)$$

Note that no composition-independent or first-order term appears in this expression. The values of the primed coefficients in Eq. (2.15) are related to the unprimed ones in Eq. (2.14) as

$$g_{0}' = g_{0} + 3g_{1} + 5g_{2} + 7g_{3}$$
$$g_{1}' = -4[g_{1} + 4g_{2} + 9g_{3}]$$
$$g_{2}' = 12[g_{2} + 5g_{3}]$$
$$g_{3}' = -32g_{3} \qquad (2.16)$$

provided that no more than four coefficients are needed, as is usually the case. The excess chemical potential of component B (the cosolvent in the case of aqueous mixtures) can similarly be written as

$$\mu_{B}^{E} = RT \ln f_{B} = g_{0}''x_{A}^{2} + g_{1}''x_{A}^{3} + g_{2}''x_{A}^{4} + g_{3}''x_{A}^{5} + \cdots \qquad (2.17)$$

where f_{B} is the activity coefficient of the cosolvent. The values of the double-primed coefficients in Eq. (2.17) are related to the unprimed ones in

TABLE 2.10 Coefficients g_j of the Excess Molar Gibbs Energies, $G^E = x(1-x)[g_0 + g_1(1-2x) + g_2(1-2x)^2 + g_3(1-2x)^3]$ J/mol, of queous Mixtures at 298.15 K, Where x is the Mole Fraction of the Cosolvent

Cosolvent	Ref.	g_0	g_1	g_2	g_3
Methanol	b	1200	−87	−330	
Ethanol	b	2907	−777	494	
2-Propanol	b	3843	−984	−98	
2-Methyl-2-propanol	d	4150	1308	879	
1,2-Ethanediol	f	−558	164	−189	
Glycerol	c	−858	−206	516	529
	c′	−508	−372	−254	
Tetrahydrofuran	e	5484	47	1371	
Dioxane	g	3835	−973	−421	
Acetone	i	4560	−163	1140	
Formic acid	m				
Acetic acid	l	3213	−118	−663	
Triethylamine	y	5346	−1023	1122	
Pyridine	n	2404	−1212	1873	
Acetonitrile	p	5253	−639	1316	
	p′	5464	421	578	
Formamide	a	−5099	4367	−2681	1063
N-Methylformamide	j	−970	−834	−425	948*
N,N-Dimethylformamide	h	−978	−653	222	
N-Methylpyrrolidin-2-one (30°C)	t	−487	−40	206	444
Hexamethyl phosphoric triamide	s	−4673	4185	−4270	
Dimethylsulfoxide	k	−4909	2168	−5	

*An additional term, $g_4(1-2x)^4$, with $g_4 = 1439$ is required.
[a]W Kangro, A Groeneveld. Z Phys Chem (NF) 32:110, 1962 (up to $x = 0.85$).
[b]RF Lama, BCY Lu. J Chem Eng Data 10:216, 1965.
[c]ECH To, JV Davies, M Tucker, P Westh, C Trandum, KSH Suh, Y Koga. J Solution Chem 28:1137, 1999 (up to $x = 0.4$).
[c′]Y Marcus. Phys Chem Chem Phys 2:4891, 2000.
[d]S Westmeier. Chem Technol 29:218, 1977.
[e]C Treiner, JF Bocquet, M Chemla. J Chim Phys 70:72, 1973.
[f]M Villamanan, C Gonzalez, HC van Ness. J Chem Eng Data 29:427, 1984.
[g]AL Vierk. Z Anorg Chem 261:283, 1950.
[h]M Cilense, AV Benedetti, R Vollet. Thermochim Acta 63:151, 1953.
[i]RV Orye, JM Prausnitz. Ind Eng Chem 57(5):18, 1965.
[j]NM Murthy, KV Sivakumar, E Rajagopal, SV Subramanyan. Acustica 48:341, 1981.
[k]SY Lam, RL Benoit. Can J Chem 52:718, 1974.
[l]JMTM Gieskes. Can J Chem 43:2448, 1965.
[m]NG Zarakhani, NP Vorobéva, Zh Fiz Khim 46:2426, 1972, reported the activity of water only.
[n]TC Chan, WA Van Hook. J Chem Soc Faraday Trans 1 72:583, 1976.
[p]C Treiner, P Tzias, M Chemla, GM Polotarskii. J Chem Soc Faraday Trans 1 72:2007, 1976.
[p′]AJ Easteal, LA Woolf. J Chem Thermodyn 20:701, 1988.
[s]C Jambon, R Philippe. J Chem Thermodyn 7:479, 1975.
[t]J Zielkiewicz. J Chem Eng Data 43:650, 1998; at 303.15 K.
[y]GL Bertrand, JW Larson, LG Hepler. J Phys Chem 72:4194, 1968.

TABLE 2.11A Excess Gibbs Energy of Mixing, $G^E/\text{J mol}^{-1}$l, of Mixtures of Benzene with Cosolvents, $G^E = x(1-x) \sum g_i(1-2x)^i$, Where x Is the Mole Fraction of the Cosolvent

Cosolvent	Ref.	T/K	g_0	g_1	g_2	g_3	g_4
Hexane	S-83	298.15	1538	285.6	75.9	41.6	
Cyclohexane							
Toluene							
Chloroform							
Tetrachloromethane	a	313.15	322.5	0	−4.0		
Methanol							
Ethanol	b	313.15	4771	−891	918		
Ethyl ether							
Tetrahydrofuran							
Dioxane							
Acetone							
Ethyl acetate							
Pyridine							
Nitromethane	b	323.15	3071	179	192		
Nitrobenzene							
Acetonitrile	b	323.15	2697	−28	292		
Dimethylformamide							
Dimethylsulfoxide							

Refs. can be found on pg. 55.

Eq. (2.14), provided again that no more than four coefficients are needed, as is usually the case, as

$$g_0'' = g_0 - 3g_1 + 5g_2 - 7g_3$$
$$g_1'' = 4[g_1 - 4g_2 + 9g_3]$$
$$g_2'' = 12[g_2 - 5g_3]$$
$$g_3'' = 32g_3 \qquad (2.18)$$

The activity of component B is $a_B = x_B f_B$ and its vapor pressure is $p_B = a_B p_B^\circ$, where p_B° is the vapor pressure of the pure component B at the given temperature. As is seen, the coefficients g_i'' in Eq. (2.18) are related to the coefficients g_i' in Eq. (2.16) due to the requirement of the Gibbs–Duhem relation (2.11).

The necessary condition for solvents to be completely miscible with each other is that

$$(\partial^2 G^E/\partial x_i^2)_{T,P} + RT/x_i(1-x_i) > 0 \qquad (2.19)$$

TABLE 2.11B Excess Gibbs Energy of Mixing, $G^E/J\,mol^{-1}$, of Mixtures of Tetrachloromethane with Cosolvents, $G^E = x(1-x)\sum g_i(1-2x)^i$, Where x is the Mole Fraction of the Cosolvent

Cosolvent	Ref.	T/K	g_0	g_1	g_2	g_3	g_4
Hexane	c	298.15	575.6	126.9			
Cyclohexane	S-79	343.15	231.8	18.7	16.8		
Benzene	a	313.15	322.5	0	-4.0		
Toluene							
Chloroform	j	298.15	93				
Methanol							
Ethanol							
Ethyl ether							
Tetrahydrofuran							
Dioxane							
Acetone							
Ethyl acetate							
Pyridine							
Nitromethane	k	323.15	5337	-469	751		
Nitrobenzene							
Acetonitrile	k	323.15	4737	-672	734		
Dimethylformamide							
Dimethylsulfoxide	d	298.15	965	270	132	67	168

TABLE 2.11C Excess Gibbs Energy of Mixing, $G^E/J\,mol^{-1}$, of Mixtures of Methanol with Cosolvents, $G^E = x(1-x)\sum g_i(1-2x)^i$, Where x Is The Mole Fraction of the Cosolvent

Cosolvent	Ref.	T/K	g_0	g_1	g_2	g_3	g_4
Hexane		298.15			Miscibility gap		
Cyclohexane		298.15			Miscibility gap		
Benzene	e	313.15	4876	465	515	271	204
Toluene	e	313.15	5196	396	507	323	834
Chloroform	S-78	323.15	3494	-1226	220	-221	
Tetrachloromethane							
Ethanol	e	313.15	-17	-15			
Ethyl ether							
Tetrahydrofuran							
Dioxane							
Acetone	S-78	323.15	1725	37			
Ethyl acetate							
Pyridine	e	313.15	-274	200	88		
Nitromethane	S-83	298.15	4160	-202	392	-88	
Nitrobenzene							
Acetonitrile							
Dimethylformamide	f	313.15	-1264	67			
Dimethylsulfoxide	d	298.15	-776	-69			

Refs. can be found on pg. 55.

TABLE 2.11D Excess Gibbs Energy of Mixing, $G^E/J\,mol^{-1}$, of Mixtures of Ethanol with Cosolvents, $G^E = x(1-x)\sum g_i(1-2x)^i$, Where x Is the Mole Fraction of the Cosolvent

Cosolvent	Ref.	T/K	g_0	g_1	g_2	g_3	g_4
Hexane	e	313.15	5424	557	932	369	173
Cyclohexane	g	298.15	5435	599	939	1306	30
Benzene	b	313.15	4771	891	918		
Toluene	e	313.15	4583	642	546	499	469
Tetrachloromethane							
Chloroform	S-78	303.15	2714	−1202	74	−320	
Methanol	e	313.15	−17	15			
Ethyl ether							
Tetrahydrofuran							
Dioxane	S-83	323.15	2474	−255	106		
Acetone	S-78	303.15	1890	103	71		
Ethyl acetate							
Pyridine	e	313.15	−39	176			
Nitromethane	S-83	298.15	5060	−147	526	−116	
Nitrobenzene							
Acetonitrile	S-82	293.15	3390	81	322	71	
Dimethylformamide	h	313.15	−506	85	34	27	
Dimethylsulfoxide							

Refs. can be found on pg. 55.

or equivalently

$$(\partial f_i/\partial x_i)_{T,P} > 0 \tag{2.20}$$

where i in the derivatives can be either A or B, over the entire composition range. This is the case for all the aqueous cosolvents listed in Table 2.10 at 298.15 K, except for triethylamine, where it applies only below the (lower) consolute temperature of 292 K. Above this temperature the mixture splits into two liquid phases. For the nonaqueous mixtures the condition of miscibility is valid for all the mixtures considered here at 298.15 K, except for those of methanol, acetonitrile, and nitromethane with n-hexane and with c-hexane. Generally, liquid–liquid equilibrium (LLE) data, i.e., the compositions of the two liquid phases at equilibrium at the given temperature, are essentially outside the scope of this book on solvent mixtures.

A certain class of binary solvent mixtures is called *regular* solutions or mixtures. These have been so coined by Hildebrand and Wood [8]

TABLE 2.11E Excess Gibbs Energy of Mixing, $G^E/\text{J mol}^{-1}$, of Mixtures of Acetone with Cosolvents, $G^E = x(1-x)\sum g_i(1-2x)^i$, Where x is the Mole Fraction of the Cosolvent

Cosolvent	Ref.	T/K	g_0	g_1	g_2	g_3	g_4
Hexane	e	313.15	4087	−10	384	−85	
Cyclohexane	g	298.15	4379	−449	−1456	240	1880
Benzene	e	313.15	1148	181	97		
Toluene	e	313.15	1536	125	128		
Tetrachloromethane							
Chloroform	S-78	323.15	−2036	383	180		
Methanol	S-78	323.15	1725	−37			
Ethanol	S-78	303.15	1890	−103	71		
Ethyl ether							
Tetrahydrofuran							
Dioxane							
Ethyl acetate							
Pyridine	S-76	303.15	588	−2	76		
Nitromethane	S-83	298.15	−258.8	−151.7	57.2		
Nitrobenzene							
Acetonitrile							
Dimethylformamide							
Dimethylsulfoxide							

Refs. can be found on pg. 55.

and by Scatchard [9] and have the following simplifying features. Only central forces between the molecules are operative, and these are pairwise additive. The mixing of the molecules is assumed to be random, so that the entropy of mixing is assumed to be ideal, i.e., $\Delta S_{mix} = -R[x_A \ln x_A + x_B \ln x_B]$, and there is no volume change on mixing. Consequently, the internal energy, enthalpy, Helmholtz energy, and Gibbs energy are all equal, and their excess values are given by the first term only in the Redlich–Kister expression (2.14), (with all $g_{j>0} = 0$):

$$U^E = H^E = A^E = G^E = x_B(1-x_B)g_0 \qquad (2.21)$$

As corollaries, for regular binary mixtures: $\mu_A^E = RT \ln f_A = g_0 x_B^2$ and $\mu_B^E = RT \ln f_B = g_0 x_A^2$. A further assumption introduced by Hildebrand and Wood [8] is that in regular solutions, where the attractive interactions between the molecules are due to dispersion forces, the

TABLE 2.11F Excess Gibbs Energy of Mixing, $G^E/\text{J mol}^{-1}$, of Mixtures of Acetonitrile with Cosolvents, $G^E = x(1-x) \sum g_i(1-2x)^i$, Where x Is the Mole Fraction of the Cosolvent

Cosolvent	Ref.	T/K	g_0	g_1	g_2	g_3	g_4
Hexane		298.15		Miscibility gap			
Cyclohexane		298.15		Miscibility gap			
Benzene	1	323.15	2697	28	292		
Toluene	S-73	318.15	3155	−134	369		
Tetrachloromethane	1	323.15	4737	672	734		
Chloroform							
Methanol							
Ethanol	S-82	293.15	3390	−81	322	−71	
Ethyl ether							
Tetrahydrofuran							
Dioxane							
Acetone							
Ethyl acetate							
Pyridine							
Nitromethane	S-83	298.15	18.6	−81.5	62.9		
Nitrobenzene							
Dimethylformamide							
Dimethylsulfoxide							

Refs. can be found on pg. 55.

internal energy can be written in terms of the *solubility parameters* δ of the components and their volume fractions:

$$U^E = \phi_A \phi_B V (\delta_A - \delta_B)^2 \tag{2.22}$$

and the same expression is also valid for G^E etc. Perusal of Tables 2.10 and 2.11 shows that none of the mixtures dealt with in the present book can be considered regular. However, if $(\delta_A - \delta_B)^2$ in Eq. (2.22) is replaced by A_{AB}, which may be treated as an empirical fitting parameter, this single fitting parameter provides a good first approximation of $G^E \sim H^E \sim \phi_A \phi_B V A_{AB}$ for a number of systems.

There are numerous other expressions (beyond the Redlich–Kister and Margules equations already presented) that have been proposed to describe the excess Gibbs energy of mixing and the activity coefficients of binary solvent mixtures, that is, to fit vapor–liquid equilibria (VLE). Some of these expressions are readily extendable to multicomponent

TABLE 2.11G   Excess Gibbs Energy of Mixing, $G^E/\mathrm{J\,mol^{-1}}$, of Mixtures of Nitromethane with Cosolvents, $G^E = x(1-x) \sum g_i(1-2x)^i$, Where x Is the Mole Fraction of the Cosolvent

Cosolvent	Ref.	T/K	g_0	g_1	g_2	g_3	g_4
Hexane		298.15		Miscibility gap			
Cyclohexane	i	318.15	5653	−352	610	−389	278
Benzene	m	323.15	3071	−179	192		
Toluene							
Tetrachloromethane	m	323.15	5337	469	751		
Chloroform							
Methanol	S-83	298.15	4160	202	392	88	
Ethanol	S-83	298.15	5060	147	526	116	
Ethyl ether							
Tetrahydrofuran							
Dioxane							
Acetone	S-83	298.15	−258.8	151.7	57.2		
Ethyl acetate	S-83	298.15	717.4	334.1	111.1		
Pyridine							
Nitrobenzene							
Acetonitrile	S-83	298.15	18.6	81.5	62.9		
Dimethylformamide							
Dimethylsulfoxide							

[a] E Puricel. Rev Roum Chim 44:223, 1999.
[b] I Brown, W Fock, F Smith. Aust J Chem 9:364, 1956.
[c] TG Bissel, AG Williamson. J Chem Thermodyn 7:131, 1975.
[d] K Quitzsch, H-P Prinz, K Sühnel, VS Pham, G Geiseler. Z Phys Chem 241:273 1969.
[e] G Kolasinska, P Oracz. Pol J Chem 55:155, 1981.
[f] J Zielkiewicz, P Oracz. Fluid Phase Equil 59:279, 1990.
[g] M Kato. Bull Chem Soc Jpn 55:23, 1982.
[h] J Zielkiewicz, A Konitz. Fluid Phase Equil 63:129, 1991.
[i] KN Marsh, HT French, HP Rogers. J Chem Thermodyn 11:897, 1979.
[j] DS Adcock, ML McGlashan. Proc R Soc A226:266, 1954.
[k] I Brown, F Smith. Aust J Chem 7:264, 1954; 8:501, 1955.
[l] I Brown, W Fock. Aust J Chem 9:180, 1956.
[m] I Brown, F Smith. Aust J Chem 8:62, 1955.
[S-nn] International Data Series, Selected Data Mixtures, Ser A, 19nn.

mixtures; see Chapter 6. Of these, the two-parameter *Wilson expression* [10], based in principle on the Flory–Huggins expression for athermal mixtures, is widely used:

$$G^E/RT = -x_A \ln(x_A + \Lambda_{AB}x_B) - x_B \ln(x_B + \Lambda_{BA}x_A) \qquad (2.23)$$

with two positive unequal parameters Λ_{BA} and Λ_{AB}. These parameters can be related to the three interaction energies between molecules of A and B, $e_{AB} = e_{BA}, e_{AA}$, and e_{BB} and to their (partial) molar volumes, V_A and V_B, by

$$\Lambda_{BA} = 1 - (V_B/V_A) \exp[-(e_{AB} - e_{AA})/k_B T] \qquad (2.24a)$$

$$\Lambda_{AB} = 1 - (V_A/V_B) \exp[-(e_{BA} - e_{BB})/k_B T] \qquad (2.24b)$$

Wilson's expression is inherently incapable of dealing with liquid–liquid immiscibility (LLE) but is sufficiently accurate for not too skew $G^E(x)$ curves.

On the whole, Wilson's expression was designed for the fitting of experimental G^E data and not for their prediction. Only two interaction parameters Λ that can be derived by fitting the experimental data are employed; therefore, the three independent interaction energies e that are of interest cannot be deduced from them individually. However, the self-interaction energies e_{AA} and e_{BB} can be derived from the Lennard–Jones expression for the potential energy of interaction of molecules of the pure components:

$$u^{LJ}(r) = 4e[(\sigma/r)^{12} - (\sigma/r)^6] \qquad (2.25)$$

which applies well for nonpolar fluids. The distance r pertains to the mass centers of particles of the fluid, the diameter of which is σ. The first term in the square brackets describes the potential energy due to repulsion at very close distances and the second term that for the attraction between the particles. The potential energy $u(r)$ can be obtained from experimental data on the pure fluids, e.g., the second virial coefficient. An estimate of the mutual interaction energy between particles of components A and B can be obtained by invoking Berthelot's geometric mean assumption: $e_{AB} = (e_{AA} \cdot e_{BB})^{1/2}$. In principle, therefore, the parameters Λ can be estimated from Eqs. (2.24) and these interaction energies in order to derive the excess molar Gibbs energy of mixing according to Wilson's expression (2.23). This would then be applicable only to mixtures of fluids, the molecules of which are distributed randomly and obey the Lennard–Jones and Berthelot expressions for the intermolecular potential.

The nonrandom two-liquid (NRTL) expression of Renon and Prausnitz [11] is based on the two-liquid approach of Scott [12] and has some analogy with the quasi-chemical approach of Guggenheim [13]. Rather than using volume fractions and interaction potential energies, as was implicitly done by Wilson, Renon and Prausnitz used mole fractions and (molar) Gibbs energies of interaction as well as a parameter descriptive of the nonrandomness of the molecular distribution. Their three-parameter expression addresses the previously mentioned drawbacks of Wilson's expression:

$$G^E/RT = x_A x_B[\tau_{BA} \exp(-\alpha\tau_{BA})/\{x_A + x_B \exp(-\alpha\tau_{BA})\}$$
$$+ \tau_{AB} \exp(-\alpha\tau_{AB})/\{x_B + x_A \exp(-\alpha\tau_{AB})\}] \qquad (2.26)$$

where $\tau_{BA} = (g_{BA} - g_{AA})/RT$, $\tau_{AB} = (g_{AB} - g_{BB})/RT$, and the g's are pairwise interaction Gibbs energies between adjacent molecules of the specified components, with $g_{BA} = g_{AB}$ (but $\tau_{BA} \neq \tau_{AB}$). The parameter α expresses the nonrandomness of the distribution of the molecules in the mixtures; it has a typical value of 0.30 for systems with moderate or low excess Gibbs energies but may vary between 0.20 and 0.55, and even higher values are permissible when fitting VLE data. On the other hand, LLE data can be fitted as well by the NRTL equation, provided $\alpha \leq 0.426$ can be used in the fitting. For good data, the probable errors in $\tau_{AB} + \tau_{BA}$ are $10/T$ and in $|\tau_{AB} - \tau_{BA}|$ they are $50/T$. These parameters are relatively weak functions of the temperature, but when obtained at a single temperature, it is impossible to obtain H^E values (see later) from them.

A widely used, but more complicated, expression for the fitting of VLE data is the universal quasi-chemical (UNIQUAC) one [14], based on Guggenheim's [13] quasi-chemical approach. It employs a quasi-lattice parameter $Z = 10$ and assigns van der Waals molecular volumes r and molecular surface areas q to the pure components. In analogy with the NRTL expression, it also uses two interaction energy parameters $\Delta u_{BA} = (u_{BA} - u_{AA})$ and $\Delta u_{AB} = (u_{AB} - u_{BB})$, being differences between three independent interaction potential energies, $u_{BA} = u_{AB}, u_{AA}$, and u_{BB}. It takes the excess Gibbs energy of mixing to be made up of the sum of a combinatorial (G^E_{comb}/RT) and an interaction (G^E_{inter}/RT) contribution:

$$G^E_{comb}/RT = x_A \ln(\phi_A/x_A) + x_B \ln(\phi_B/x_B) + (Z/2)[q_A x_A \ln(\theta_A/\phi_A)$$
$$+ q_B x_B \ln(\theta_B/\phi_B)] \qquad (2.27)$$

$$G^E_{inter}/RT = -q_A x_A \ln[\theta_A + \theta_B \exp(-\Delta u_{BA}/RT)]$$
$$- q_B x_B \ln[\theta_B + \theta_A \exp(-\Delta_{AB}/RT)] \qquad (2.28)$$

Here the quantities ϕ_i are the volume fractions defined as $x_i r_i/(x_A r_A + x_B r_B)$ and θ_i are the corresponding surface fractions: $x_i q_i/(x_A q_A + x_B q_B)$, with $i =$ A and B. Value for r and q of the solvents considered here are listed in Table 2.12. It has been stressed by Luecke [15] that the values of

TABLE 2.12   UNIQUAC Molecular Volumes r and Surface Areas q of Solvents

Solvent	r	q
Hexane	4.500	3.856
c-Hexane	4.046	3.240
Benzene	3.188	2.400
Toluene	3.923	2.968
Water	0.920	1.400
Methanol	1.431	1.432
Ethanol	2.576	2.588
2-Propanol	3.249	3.124
2-Methyl-2-propanol (*t*-butanol)	3.923	3.744
Ethylene glycol	2.409	2.248
Glycerol	3.856	3.676
Diethyl ether	3.395	3.016
Tetrahydrofuran	2.942	2.400
1,4-Dioxane	3.185	3.860
Acetone	2.574	2.336
Formic acid	1.528	1.532
Acetic acid	2.202	2.072
Ethyl acetate	3.723	3.356
Chloroform	2.870	2.410
Tetrachloromethane	3.390	2.910
Triethylamine	5.012	4.256
Pyridine	2.999	2.113
Nitromethane	2.009	1.868
Nitrobenzene	4.076	3.104
Acetonitrile	1.870	1.724
Formamide	1.466	1.336
N-Methlformamide	2.367	2.184
N,N-Dimethylformamide	3.086	2.736
N-Methylpyrrolidin-2-one	4.390	3.804
Hexamethyl phosphoric triamide	7.285	6140
Dimethylsulfoxide	2.827	2.472

Δu_{BA} and Δu_{AB} are mutually correlated to an appreciable extent, hence are not unique for fitting the $G^E(x)$ data. Therefore, the pairwise interaction (potential) energies u *cannot* be deduced from the Δu fitting parameters. A simplification of the UNIQUAC calculation was introduced by Abrams and Prausnitz [14] by the assumption that $u_{AA} = -(\Delta_{vap}H_A - RT)/q_A$ and $u_{BB} = -(\Delta_{vap}H_B - RT)/q_B$ and that $u_{AB} = u_{BA} =$

$(u_{AA}u_{BB})^{1/2}(1 - c_{AB})$. This approximation uses the enthalpies of vaporization of the pure components, $\Delta_{vap}H_i$, and a single mixing parameter, c_{AB}, that is positive and small compared with unity for mixtures of nonpolar components. When one or both of the components are polar, larger values of c_{AB} must be taken into account, e.g., 0.494 for tetrachloromethane + acetonitrile at 45°C, or even negative values, e.g., −0.183 for acetone + chloroform at 50°C.

There still remains the question of whether the VLE can be *predicted* for binary (or higher multicomponent) mixtures from known properties of the pure components at the ambiance in question. Once the function $G^E(x)$ is known, the partial vapor pressures and vapor composition can be obtained from Eqs. (2.14) to (2.18). Attempts to solve this problem have been made by invoking the additivity of group contributions in the analytical solution of group (ASOG) [16] and the UNIFAC [17] methods. The latter employs the UNIQUAC approach, utilizing the combinatorial part, Eq. (2.27), directly, as it requires only pure component properties and deals adequately with the size differences of the molecules of the components. The interaction part, Eq. (2.28), however, requires consideration of the interactions of various functional groups in the molecules. These groups are characterized by volume and area parameters, summed to yield the r and q values for the molecules, and mutual interaction parameters. There are more than 40 main groups subdivided into nearly 90 subgroups (e.g., CH_3CO- is a different subgroup from $-CH_2CO-$), so that altogether hundreds of interaction parameters are needed to cover all the eventualities. Their consideration is outside the scope of this book, but see Ref. 18.

In principle, once $G^E(x)$ is known as a function of the pressure and the temperature, the derivative quantities, such as the excess volume and entropy of mixing, can be calculated from the partial derivatives $(\partial G^E(x)/\partial P)_{x,T}$ and $(\partial G^E(x)/\partial T)_{x,P}$, etc. In practice, it is generally more accurate to obtain such functions by direct measurements of the thermodynamic properties, i.e., densities, calorimetric heats of mixing, etc.

Ideal mixing of the solvents produces no heat of mixing; therefore, any heat evolved or absorbed during mixing is an *excess enthalpy of mixing*, H^E. Such heats are conveniently and accurately measure calorimetrically and are generally reported to ±1 J mol^{-1} or better. The calorimeters for this purpose are designed so that the vapor space is minimal in order to avoid corrections for the heats of vaporization. In principle, the van't Hoff relation may be applied to the excess Gibbs energies of mixing, ultimately to vapor pressure and vapor composition measurements at several temperatures at a given external pressure, P:

$$H^E = -T^2[\partial(G^E/T)/\partial T]_P \tag{2.29}$$

However, the derivatization diminishes the accuracy considerably, and this procedure should be avoided if calorimetric data can be obtained. The procedure can be inverted by integration, however, and the excess Gibbs energy of mixing can be calculated at a temperature T_2 from the excess heat of mixing, if its value at another temperature, T_1, is known.

$$G^E(T_2)=G^E(T_1)-T\int_{T_1}^{T_2} H^E d(1/T)\approx(T_2/T_1)G^E(T_1)+H^E(1-T_2/T_1) \tag{2.30}$$

The approximation ignores the temperature dependence of H^E over a short temperature interval. This procedure is rather useful because vapor pressures and vapor compositions of solvent mixtures are often measured at a relatively high temperature, say 40 to 60°C (313 to 333 K), where the volatilities of the components are higher and the vapor pressures more readily measurable accurately than at the lower temperature, say 25°C (298 K), where values of G^E are required.

The heats of mixing in binary solvent mixtures can also be written in terms of Redlich–Kister-type expressions:

$$H^E = x_B(1-x_B)[h_0 + h_1(1-2x_B) + h_2(1-2x_B)^2 + h_3(1-2x_B)^3 + \cdots] \tag{2.31}$$

The coefficients h_j of Eq. (2.31) for 298.15 K are given in Table 2.13 for the 20 common cosolvent mixtures with water. The corresponding coefficients for the nonaqueous mixtures are given in Table 2.14. The heats of mixing curves, $H^E(x_B)$, tend to be more skewed than the excess Gibbs energy curves, $G^E(x_B)$. In fact, the $H^E(x_B)$ curves for many systems are S-shaped and have positive and negative segments. This is manifested by the odd-index coefficients (responsible for the skewness) of H^E being larger than the corresponding ones of G^E. The partial molar enthalpies of the components, h_A^E and h_B^E, are expressed by the replacement of the symbols μ^E in Eqs. (2.14) and (2.17) by h^E and are calculated by using the h_j parameters from Tables 2.13 and 2.14 to replace g_i in Eq. (2.15) and subsequently replacing g_i', and g_i'' in Eqs. (2.16) and (2.18) by h_i' and h_i''.

TABLE 2.13 Coefficient h_j of the Excess Molar Enthalpies, $H^E = x(1-x)$ $[h_0 + h_1(1-2x) + h_2(1-2x)^2 + h_3(1-2x)^3]$ J/mol, of Aqueous Mixtures at 298.15 K, where x Is the Mole Fraction of the Cosolvent

Cosolvent	Ref.	h_0	h_1	h_2	h_3
Methanol	b	−3102	2040	−2213	
Ethanol	b	−1300	−3567	−4971	
2-Propanol	b	854	5167	−7243	
2-Methyl-2-propanol	c	170	−4129	1136	4653*
1,2-Ethanediol	d	−2776	1933	−1172	
Glycerol	g	−2466	−1343	−125	
Tetrahydrofuran	f	−136	7443	−4229	
	r	−434	−2710	−3629	−7044
Dioxane	h	611	6006	−1712	
Acetone	i	569	−5408	−1838	
Formic acid	n	−1027	−500	−3788	
Acetic acid	j	1239	1791	310	
Triethylamine	o	−8294	7834	−4801	
Pyridine	k	−5600	−1020	1756	
Acetonitrile	l	4640	2922	−1028	−8
	m	4157	−3075		
Formamide	a	1074	152	381	273
N-Methylformamide	e†	−3635	−2438	−529	−108
N,N-Dimethylformamide	d	−7616	7751	−1904	
N-Methylpyrrolidin-2-one	o	−9983	8917	−2496	
Hexamethyl phosphoric triamide	p	−11367	−10085	−11288	
Dimethylsulfoxide	q	−10372	6922	−2466	

*Additional terms required, with $h_4 = -7698$ and $h_5 = -13530$.

†At 308.15 K.

[a] AM Zaichikov, EA Nogovitsyn, NI Zhelesniak, GA Krestov. Zh Fiz Khim 62:3118, 1988; Russ J Phys Chem 62:1630, 1988.

[b] RF Lama, BCY Lu. J Chem Eng Data 10:216, 1965.

[c] S Westmeier. Chem Technol 29:218, 1977.

[d] K Rehm, HJ Bittrich. Z Phys Chem 251:109, 1972.

[e] AM Zaichikov, OE Golubinskii. Zh Fiz Khim 70:1175, 1996; Russ J Phys Chem 70:1090, 1996.

[f] DN Glew, H Watts. Can J Chem 51:1933, 1973.

[g] Y Marcus. Phys Chem Chem Phys 2:4891, 2000.

[h] KW Morkom, RW Smith. Trans Faraday Soc 66:1073, 1970.

[i] DO Hanson, M Van Winkle. J Chem Eng Data 5:30, 1960.

[k] JMTM Gieskes. Can J Chem 43:2448, 1965.

[l] TC Chan, WA Van Hook. J Chem Soc Faraday Trans 1 72:583:1976.

[m] C Treiner, P Tzias, M Chemla, GM Polotarskii. J Chem Soc Faraday Trans 1 72:2007, 1976.

[n] AJ Easteal, LA Woolf. J Chem Thermodyn 20:701, 1988.

[o] AN Campbell, AJR Campbell. Trans Faraday Soc 29:1240, 1933.

[p] GL Bertrand, JW Larson, LG Hepler. J Phys Chem 72:4194, 1968.

[q] MF Fox, KP Wittingham. J Chem Soc Faraday Trans 1 71:1407, 1975.

[r] DV Batov, AM Zaichikov, VP Slyusar, VP Korolev. Russ J Gen Chem 69:1866, 1999.

TABLE 2.14A   Excess Enthalpy of Mixing, $H^E/\text{J mol}^{-1}$, of Mixtures of Benzene with Cosolvents, $H^E = x(1 - x) \Sigma\, h_i(1 - 2x)^i$, Where x Is the Mole Fraction of the Cosolvent

Cosolvent	Ref.	T/K	h_0	h_1	h_2	h_3	h_4
Hexane	S-76	323.15	3244	567	128	59	−25
Cyclohexane	S-74	308.15	3123	137	93	56	
Toluene	S-82	298.15	272.0	20.9			
Chloroform	a	298.15	−1690	330	−85		
Tetrachloromethane	S-73	298.15	462.2	15.9	33.2	6.4	
Methanol	S-76	298.15	2516	1112	285	3104	3671
Ethanol	S-76	298.15	3076	1635	240	2993	3728
Ethyl ether	b	298.15	−33.9	232.2	48.1		
Tetrahydrofuran							
Dioxane	S-73	298.15	−129	−254	6	−40	−77
Acetone	c	303.15	576	82	193		
Ethyl acetate	d	298.15	394	−104	−16	120	
Pyridine	S-73	298.15	32.7	26.4	12.0		
Nitromethane	e	298.15	3340	824			
Nitrobenzene	S-74	293.15	1048	−178	−94		
Acetonitrile	S-83	298.15	1777	−617	234		
Dimethylformamide	f	298.15	74	−30	84	237	268
Dimethylsulfoxide	S-74	298.15	2343	100	586		

Refs. can be found on pgs. 68 and 69.

The *heat of solution* of component B in the binary mixtures, $\Delta_{\text{solution}}H_B$, is related to its partial molar enthalpy:

$$\Delta_{\text{solution}}H_B = h_B/(1 - x_B) \tag{2.32}$$

At infinite dilution, the limiting heat of solution is

$$\Delta_{\text{solution}}H_B^{\infty} = h_B^{\infty} = \sum h_j \tag{2.33}$$

i.e., the sum of the coefficients of Eq. (2.31). Conversely, the heat of solution of component A in the mixtures is $\Delta_{\text{solution}}H_A = h_A/x_B$, which becomes $\Delta_{\text{solution}}H_A^{\infty} = h_0 - h_1 + h_2 - h_3$ for infinitely dilute dissolution of it in neat component B. The *relative* partial molar enthalpy of component B, $l_B = h_B - h_B^{\circ} = h_B^E$ is the excess partial molar enthalpy of this component, obtained from Eq. (2.17) by replacement of g'' by h'', and similarly for l_A and h_A^E, using the correspondingly modified Eq. (2.15).

The excess entropies of solvent mixtures are given by the identity $S^E \equiv (H^E - G^E)/T$. The Redlich–Kister expression for the excess

TABLE 2.14B Excess Enthalpy of Mixing, $H^E/J\,mol^{-1}$ of Mixtures of Tetrachloromethane with Cosolvents, $H^E = x(1-x) \Sigma\, h_i(1-2x)^i$, Where x Is the Mole Fraction of the Cosolvent

Cosolvent	Ref.	T/K	h_0	h_1	h_2	h_3	h_4
Hexane	S-75	298.15	1264	−195	56	53	
Cyclohexane	S-73	298.15	664.2	29.4	12.7		
Benzene	S-73	298.15	462.2	−15.9	33.2	−6.4	
Toluene							
Chloroform	g	298.15	593	4			
Methanol	h	298.15	782	1487	−570	−1436	3253*
Ethanol							
Ethyl ether	S-73	298.15	−1973	−39	502	62	−870
Tetrahydrofuran							
Dioxane	S-74	298.15	−1002	62	157	42	
Acetone	i	323.15	1031	1046	680		
Ethyl acetate							
Pyridine	j	298.15	−451	963	321	38	−272
Nitromethane	e	298.15	5266	528	1950		
Nitrobenzene							
Acetonitrile	S-83	298.15	3089	−473	1155	−917	
Dimethylformamide							
Dimethylsulfoxide	S-74	298.15	723	923	1089	860	

*A term in $h_5 = 4791$ is also required.
Refs. can be found on pgs. 68 and 69.

entropy is obtained by using the corresponding terms from the excess enthalpy and Gibbs energy of mixing, the coefficients being $s_j \equiv (h_j - g_j)/T$. Given the entries for $T = 298.15\,K$ in Tables 2.10 and 2.13 for the common aqueous mixtures and in Tables 2.11 and 2.14 for the nonaqueous mixtures, these coefficients need not be reported here. The excess entropy curves, $S^E(x_B)$ have a skewness similar to that of the excess enthalpy curves, $H^E(x_B)$, which fact makes the excess Gibbs energy curves, $G^E(x_B)$, much more symmetrical than either. The partial molar entropies are calculated in an analogous manner to the chemical potentials or partial molar enthalpies of mixing.

Typical curves of $G^E(x_B)$, $H^E(x_B)$, and $TS^E(x_B)$ are shown in Fig 2.3a–d. The water + dimethylsulfoxide system (a) has all three functions negative, therefore $G^E(x_B)$ is less negative than $H^E(x_B)$, but mutual attraction of the molecules of the two components is indicated. The water + ethanol system (b) also has negative $H^E(x_B)$ and $TS^E(x_B)$,

TABLE 2.14C Excess Enthalpy of Mixing, $H^E/\text{J mol}^{-1}$ of Mixtures of Methanol with Cosolvents, $H^E = x(1-x) \Sigma\, h_i(1-2x)^i$, Where x Is the Mole Fraction of the Cosolvent

Cosolvent	Ref.	T/K	h_0	h_1	h_2	h_3	h_4
Hexane*	S-76	298.15	3658	1223	−4430	−2960	7741
Cyclohexane		298.15			Miscibility gap		
Benzene	S-76	298.15	2516	−1112	285	−3104	3671
Toluene	S-76	298.15	2515	−1061	330	−3045	3980
Chloroform	h	298.15	−1199	−4019	−302	−2256	3446
Tetrachloromethane	h	298.15	782	−1487	−570	1436	3253[†]
Ethanol	S-76	298.15	17.9	8.0	−0.8	−3.2	
Ethyl ether	k	298.15	−204	1508	1010	−3648	2048
Tetrahydrofuran	l	298.15	2127	−933	83	−209	
Dioxane	m	303.15	4160	−500	−1710		
Acetone	d	298.2	2724	−427	16.6		
Ethyl acetate							
Pyridine	m	303.15	−3360	−899	1521		
Nitromethane	n	298.15	4736	−2557	2399	−2011	
Nitrobenzene	n	298.15	3324	−2175	2131	−6415	5521
Acetonitrile	h	298.15	4345	827	781	231	34
Dimethylformamide	o	293.15	−428	47	−126		
Dimethylsulfoxide	p	298.15	−1564	242	501	33	

*Outside the range $0.211 \le x \le 0.736$; inside this range phase separation takes place and $H^E = 464 + 73x$ J mol^{-1}.
[†]A term in $h_5 = -4791$ is also required.
Refs. can be found on pgs. 68 and 69.

but because the latter is more negative than the former, the resulting $G^E(x_B) = H^E(x_B) - TS^E(x_B)$ is positive. Note also that $G^E(x_B)$ is more symmetrical for this system than the other functions. The ethanol + nitromethane system (c) has all three functions positive, but because $TS^E(x_B) < H^E(x_B)$, the Gibbs energy is not as positive as the enthalpy. Here the self-interactions of the ethanol molecules by hydrogen bonding are responsible for the nonideality and nitromethane, being only a weak hydrogen bond acceptor, is rejected from the vicinity of the ethanol. The maximal $G^E(x_B)$ is not much below the phase separation limit for such a system, with the coefficient $g_0 = 5060$ J mol$^{-1} \sim 2RT = 4958$ J mol^{-1}. Finally, the system methanol + ethanol (d) is not far from ideality, the absolute size of the functions being quite small, and the positive

TABLE 2.14D   Excess Enthalpy of Mixing, H^E/J mol^{-1} of Mixtures of Ethanol with Cosolvents, $H^E = x(1 - x) \Sigma h_i(1 - 2x)^i$, Where x Is the Mole Fraction of the Cosolvent

Cosolvent	Ref.	T/K	h_0	h_1	h_2	h_3	h_4
Hexane	S-76	298.15	2481	−265	−521	−2787	4997
Cyclohexane	q	298.15	2547	319	455	1466	1897
Benzene	S-76	298.15	3076	−1635	240	−2993	3728
Toluene	S-76	298.15	2903	−1487	28	−3124	4239
Tetrachloromethane							
Chloroform							
Methanol	S-73	298.15	17.9	−8.0	−0.8	3.2	
Ethyl ether	k	298.15	1121	−22	1460	−2082	1321
Tetrahydrofuran	l	298.15	3163	−1025	207	313	
Dioxane	l	298.15	6134	−388	641	210	
Acetone	r	298.15	4452	−200	469		
Ethyl acetate	S-73	298.15	5121	−1025	941		
Pyridine	s	298.15	−650	−673			
Nitromethane	t	298.15	6532	−1357	1571	−191	
Nitrobenzene							
Acetonitrile	ff	298.15	6007	734	1034	201	273
Dimethylformamide	t	298.15	1612	−447	−292		
Dimethylsulfoxide	u	298.15	386	503	103	−420	

Refs. can be found on pgs. 68 and 69.

$TS^E(x_B) > H^E(x_B)$ causing a negative $G^E(x_B)$, the hydrogen bonding of the molecules of the two components being compatible with each other. A detailed consideration of the thermodynamic functions of all the other systems dealt with in Tables 2.10 and 2.13 and in Tables 2.11 and 2.14 in terms of the interactions taking place in the mixtures is outside the scope of this book. The references to these tables should provide access to this information.

The (isobaric) *heat capacity* of a binary solvent mixture *at constant pressure* is readily measurable calorimetrically. Care must be taken to minimize the vapor phase in measurements of the isobaric heat capacity in order to avoid the need to correct for the heat capacity of the vapor. This is done most conveniently by flow microcalorimetry, where the heat absorbed per unit volume and unit increase in temperature is measured and the specific volume (or density) must be measured concurrently. The accuracy of such measurements is high, often better than

TABLE 2.14E   Excess Enthalpy of Mixing, H^E/J mol^{-1} at 298.15 K of Mixtures of Acetone with Cosolvents, $H^E = x(1 - x) \Sigma h_i(1 - 2x)^i$, Where x Is the Mole Fraction of the Cosolvent

Cosolvent	Ref.	T/K	h_0	h_1	h_2	h_3	h_4
Hexane	v	298.15	6384	−836	877		
Cyclohexane	w	298.15	69301	694	1377		
Benzene	c	303.15	576	−82	193		
Toluene	S-73	318.15	1835	−32	5		
Tetrachloromethane	i	323.15	1031	−1046	680		
Chloroform	gg	298.15	−7561	2390	924	−877	
Methanol	d	298.2	2724	427	16.6		
Ethanol	r	298.15	4452	200	469		
Ethyl ether	d	298.19	2031	−380	219		
Tetrahydrofuran	x	308.15	775	−230	262	383	
Dioxane							
Ethyl acetate	y	308.15	542	−55	−133	169	
Pyridine	z	298.15	242	−310	143	285	
Nitromethane	i	323.15	−656	−282	55		
Nitrobenzene							
Acetonitrile	S-83	298.15	−432	−74	−12		
Dimethylformamide							
Dimethylsulfoxide							

Refs. can be found on pgs. 68 and 69.

± 0.1 J K^{-1} mol^{-1}. The excess molar isobaric heat capacity, C_P^E, can, again, be written as a Redlich–Kister-type expression:

$$C_P^E = x_B(1 - x_B)[c_0 + c_1(1 - 2x_B) + c_2(1 - 2x_B)^2 + c_3(1 - 2x_B)^3 + \cdots]$$
(2.34)

However, there are very few systematic reports of isobaric heat capacity data for nonaqueous mixtures, hence the data reported here are limited to aqueous mixtures. The coefficients for the common aqueous mixtures of cosolvents at 298.15 K are shown in Table 2.15. Examples of the excess heat capacities at 298.15 K of some aqueous mixtures of amides—formamide (FA), N-methylformamide (NMF) and hexamethyl phosphoric triamide (HMPT)—are shown in Fig. 2.4 [19]. The partial molar isobaric heat capacities are calculated in an analogous manner to the chemical potentials or partial molar heats of mixing.

The *isochoric heat capacity* of a liquid, C_V, i.e., that at constant volume, is of little practical importance except near the critical point,

TABLE 2.14F   Excess Enthalpy of Mixing, $H^E/$J mol^{-1} of Mixtures of Acetonitrile with Cosolvents, $H^E = x(1 - x) \Sigma h_i(1 - 2x)^i$, Where x Is the Mole Fraction of the Cosolvent

Cosolvent	Ref.	T/K	h_0	h_1	h_2	h_3	h_4
Hexane*	aa	300.05	−4937	−4050	20681	7710	
Cyclohexane[†]	bb	323.15	5675	2603	6432	−5230	279
Benzene	S-83	298.15	1777	617	234		
Toluene							
Tetrachloromethane	S-83	298.15	3089	473	1155	917	
Chloroform	S-83	298.15	−3133	1508	216	−208	
Methanol	hh	298.15	4345	−827	781	−231	34
Ethanol	hh	298.15	6007	−734	1034	−201	273
Ethyl ether	x	298.15	1599	143	77	−112	
Tetrahydrofuran	x	298.15	359	493	−308	−340	
Dioxane	x	298.15	−58	−297	210	70	
Acetone	S-83	298.15	−432	74	−12		
Ethyl acetate	S-82	298.15	160	−4	−165		
Pyridine							
Nitromethane							
Nitrobenzene	cc	298.16	1125	380	171		
Dimethylformamide	dd	298.15	−807.7	−39.6	8.8		
Dimethylsulfoxide	ee	298.15	126	121	−9		

*Outside the miscibility gap $0.155 \leq x \leq 0.868$; inside it $H^E = 824 - 118x$ J mol^{-1}
[†]At 0.4 MPa, outside the miscibility gap $0.135 \leq x \leq 0.880$; inside it $H^E = 1060.5 - 30.0x$ J mol^{-1}.
Refs. can be found on pgs. 68 and 69.

where $(\partial P/\partial T)_V$ is small, so that the pressure developed on heating is not too large. It can be calculated from

$$C_V = C_P - \alpha_P^2 VT/\kappa_T \tag{2.35}$$

and requires knowledge of the isobaric expansivity, α_P, isothermal compressibility, κ_T, and molar volume, V, of the liquid. The same applies to liquid mixtures, for which these quantities are discussed subsequently.

The volume change on mixing is expressed by means of the *excess volume of mixing*, V^E, which may be a deficiency rather than an excess (i.e., have a negative value). The excess volume of mixing is defined as

$$V^E = V - \sum_i x_i V_i = \left(\sum_i x_i M_i \right)/d - \sum_i x_i(M_i/d_i) \tag{2.36}$$

TABLE 2.14G Excess Enthalpy of Mixing, $H^E/\text{J mol}^{-1}$, of Mixtures of Nitromethane with Cosolvents, $H^E = x(1-x)\sum h_i(1-2x)^i$, Where x Is the Mole Fraction of the Cosolvent

Cosolvent	Ref.	T/K	h_0	h_1	h_2	h_3	h_4
Hexane	e	298.15	Miscibility gap : $0.026 \leq x \leq 0.964$				
Cyclohexane	e	298.15	Miscibility gap : $0.037 \leq x \leq 0.969$				
Benzene	e	298.15	3283	808	400	52	
Toluene							
Tetrachloromethane	e	298.15	5266	−116	1953	−1236	
Chloroform							
Methanol	n	298.15	4736	2557	2399	2011	
Ethanol	t	298.15	6532	1357	1571	191	
Ethyl ether							
Tetrahydrofuran							
Dioxane							
Acetone	i	323.15	−656	282	55		
Ethyl acetate							
Pyridine	z	298.15	1880	351	155	398	
Nitrobenzene							
Acetonitrile							
Dimethylformamide							
Dimethylsulfoxide							

[a]RP Rastogi, J Nath, RR Misra. J Chem Thermodyn 3:307, 1971.
[b]J -EA Otterstedt, RW Missen. J Chem Eng Data 11:360, 1966.
[c]HH Möbius. J Prakt Chem 2:95, 1955.
[d]H Hirobe, 1926, quoted by J Timmermans. The Physico-chemical Constants of Binary Systems in Concentrated Solutions. Vols. 1 and 2. New York: Interscience, 1959.
[e]KN Marsh. J Chem Thermodyn 17:29, 1985.
[f]H Ukibe, R Tanaka, S Murakami, R Fujishiro. J Chem Thermodyn 6:201, 1974.
[g]DS Adcock, ML McGlashan. Proc R Soc A226:266, 1954.
[h]I Nagata, K Tamura. Fluid Phase Equil 15:67, 1983.
[i]I Brown, W Fock. Aust J Chem 10:417, 1957.
[j]DF Gray, ID Watson, AG Willamson. Aust J Chem 21:379, 1968.
[k]MA Villamanan, C Casanova, AH Roux, J-PE Grolier. J Chem Thermodyn 14:251, 1982.
[l]TM Letcher, UP Govender. J Chem Eng Data 40:1097, 1995.
[m]PP Singh, DV Verma, PS Arora. Thermochim Acta 15:267, 1976.
[n]I Lee, CH Kang, B-S Lee, HW Lee. J Chem Soc Faraday Trans 86:1477, 1990.
[o]L Grote, G Riesel, H-J Bittrich. Z Chem 7:444, 1967.
[p]K Quitzsch, H-P Prinz, K Sühnel, VS Pham, G Geiseler. Z Phys Chem 241:273, 1969.
[q]G Conti, P Gianni, E Matteoli. Thermochim Acta 247:293, 1994.
[r]GL Nicolaides, CA Eckert. J Chem Eng Data 23:152, 1978.
[s]TJV Findlay, JL Copp. Trans Faraday Soc 65:1463, 1969.
[t]I Hammerl, A Feinbube, K Herkner, H-J Bittrich. Z Phys Chem 271:1133, 1990.

TABLE 2.14   Footnotes continued

[u]JJ Lindberg, I Pietilä. Suom Kemist 35B:30, 1962.
[v]S Murakami, K Amaya, R Fujishiro. Bull Chem Soc Jpn 37:1776, 1984.
[w]BS Lark, S Kaur, S Singh. Thermochim Acta 105:219, 1986.
[x]TM Letcher, U Domanska. J Chem Thermodyn 26:75, 1994.
[y]S Shen, Y Wang, J Shi, G Benson, BC-Y Lu. J Chem Eng Data 37:400, 1992.
[z]OK Shebanova, LV Kurits'in. Izv Vyshch Uchebn Zaved Khim Khim Tekhnol 39:122, 1996.
[aa]ML Lakhanpal, HG Mandal, SC Ahuja. Indian J Chem A20:1008, 1981.
[bb]JB Ott, JE Purdy, BJ Neely, RA Harris. J Chem Thermodyn 20:1079, 1988.
[cc]L Jannelli, M Pansini, A Lopez. Fluid Phase Equil 15:219, 1983.
[dd]S Miyanaga, K Tamura, S Murakami. J Thermal Anal 38:1767, 1992.
[ee]M Nakamura, K Chubachi, K Tamura. J Chem Thermodyn 25:1311, 1993.
[ff]I Nagata, K Tamura. J Chem Thermodyn 18:39, 1986.
[gg]I Nagata, S Ozaki, K Myohen. J Chem Thermodyn 24:607, 1992.
[hh]I Nagata, K Tamura. Fluid Phase Equil 24:289, 1985.
[s-nn]International Data Series, Selected Data Mixtures, Ser A, 19nn.

and is obtained from measurements of the density of the mixture as a function of the composition, $d(x)$, knowing the densities of the components, d_i, under the thermodynamic conditions employed. Excess volumes are commonly obtained to ± 0.01 cm^3 mol^{-1} and are described in terms of polynomial functions of the mole fractions of the components. In the case of binary mixtures of solvent A and a cosolvent (component B), a Redlich–Kister-type expression is often used. It has a form that ensures that the values of V^E are exactly zero for the composition limits of pure A and pure B:

$$V^E = x_B(1 - x_B)[v_0 + v_1(1 - 2x_B) + v_2(1 - 2x_B)^2 + v_3(1 - 2x_B)^3 + \cdots]$$

(2.37)

The coefficients v_j with even indexes j produce symmetrical curves of $V^E(x_B)$, whereas those with odd indexes produce skewness in the curves. The coefficients can be used to reconstitute the values of V^E of Eq. (2.37) or of V and d in Eq. (2.36) for given compositions. For the equimolar mixture $x_B = x_A = 0.5$, hence $V^E = v_0/4$, the other coefficients being irrelevant at this composition.

For the common binary aqueous solvent mixtures $V^E(x_A = 0.5)$ varies from ~ -0.1 for aqueous formamide to ~ -1.5 for aqueous acetone. In fact, V^E is generally negative over the entire composition range when one of the components is water. The values of the coefficients v_j for Eq. (2.37) at 298.15 K of aqueous mixtures of the 20 common cosolvents are shown in Table 2.16. Four coefficients, v_0 to v_3,

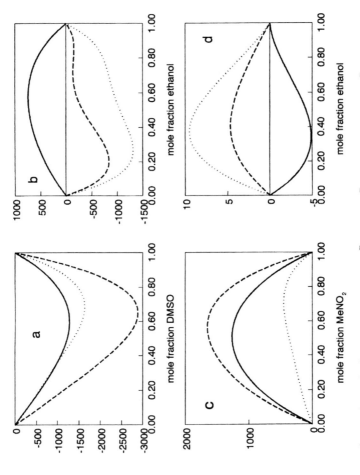

FIGURE 2.3 Excess thermodynamic functions: (———) G^E, (- - - -) H^E, and (. . . . .) TS^E at 298.15 K for the systems (a) water + dimethylsulfoxide (DMSO), (b) water + ethanol, (c) ethanol + nitromethane (MeNO₂), and (d) methanol + ethanol.

TABLE 2.15 Coefficients c_j of the Excess Molar Isobaric Heat Capacities, $C_P^E = x(1 - x)[c_0 + c_1(1 - 2x) + c_2(1 - 2x)^2 + c_3(1 - 2x)^3]$ J/K·mol, of Aqueous Mixtures at 298.15 K, Where x Is the Mole Fraction of the Cosolvent

Cosolvent	Ref.	c_0	c_1	c_2	c_3
Methanol	a	15.8	37.7	−15.5	
Ethanol	a	41.9	40.2	58.7	
2-Propanol	b	168	−587	742	
2-Methyl-2-propanol	b	296	−1166	1256	
1,2-Ethanediol	j	1.36	4.85	3.24	15.49*
Glycerol	c	49.9	−9.3		
Tetrahydrofuran	k	−22.4	−200.4	−160.7	
Dioxane	d	25.9	10.8	−11.1	66.9
Acetone	b	37.4	−40.4	117.4	
Formic acid	e	−38.5	27.0		
Acetic acid	e	−60.7	104.1		
Triethylamine					
Pyridine					
Acetonitrile	l	35.0	−8.5	8.9	35.1
Formamide	f	−4.6	−8.0	−1.1	
N-Methylformamide	f	11.3	17.1	15.9	
N,N-Dimethylformamide	f	38.1	34.0	12.2	
N-Methylpyrrolidin-2-one	g	51.3	37.2	79.9	
Hexamethyl phosphoric triamide	i	19.7	−34.2	118.3	218.7
Dimethylsulfoxide	h	−39.8	−52.9	−61.0	

*Another term, 18.05 $(1 - 2x)^4$, is required.
[a]E Bose. Z Phys Chem 58:585, 1907.
[b]R Arnaud, L Avedikian, J-P Morel. J Chim Phys 69:45, 1972.
[c]SP Sukhatme, N Saikhedkar. Indian J Technol 7:1, 1969.
[d]RD Stallart, ES Amis. J Am Chem Soc 74:1781, 1952 (40°C).
[e]CR Bury, DG Davies. J Chem Soc 1932:2413; T Ackermann, F Schreiner. Z Elektrochem 62:1143, 1958.
[f]C deVisser, G Somsen. Z Phys Chem 92:159, 1974.
[g]DD Macdonald, D Dunay, G Hanlon, JB Hyne. Can J Chem Eng 49:420, 1971.
[h]R Philippe, C Jambon. J Chim Phys 71:1041, 1974.
[i]MV Kulikov, AM Kolker, AG Krestov. Zh Priklad Khim 63:852, 1990. A better fit for $x > 0.1$ of data of these authors [AM Kolker,MV Kulikov, AG Krestov. Izv Vyssh Ucheb Zaved Khim Khim Tekhnol 28:11, 1985] is $15.3 - 7.08x - 32.33x^2 + 23.87x^3$.
[j]G Douheret, A Pal, H Høiland, O Anowi, MI Davis J Chem Thermodyn 23:569, 1991.
[k]M Costas, D Patterson. J Chem Soc Faraday Trans 1 81:2381, 1985; data only at $x \geq 0.76$.
[l]GC Benson, PJ D'Arcy, YP Handa. Thermochim Acta 46:295, 1981; C DeVisser, WJM Heuvelsland, LA Dunn, G Somsen. J Chem Soc Faraday Trans 1 74:1159, 1978—mean of the two sets, differing by ±5% from each set.

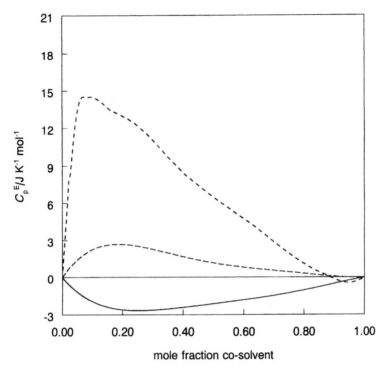

FIGURE 2.4 Excess heat capacity at 298.15 K of aqueous FA (———), NMF (— —), and DMF (- - - - -).

generally suffice for the description of V^E within its experimental accuracy, although in some cases up to six coefficients have to be used. Examples of the function $V^E/x(1 - x)$ are shown in Fig. 2.5 and it is seen from the curvature that more than two coefficients v_j are needed (the function would be linear with two coefficients only). The coefficients for the nonaqueous mixtures are shown in Table 2.17. Again, four or maximally five coefficients v_j are sufficient to describe the V^E data adequately. The excess molar volumes of some nonaqueous mixtures are positive and of some negative, and considerable skewness of the $V^E(x_B)$ curves and eventual changes of sign are common. The coefficients with even indexes are the same in tables of mixtures of solvent A with various solvents B as those in tables of solvent B with various solvents A when they pertain to the same pair of solvents. This is not the case of the coefficients with odd indexes: these change sign between such tables.

TABLE 2.16 Coefficients v_j of the Excess Molar Volumes, $V^E = x(1 - x)$ $[v_0 + v_1(1 - 2x) + v_2(1 - 2x)^2 + v_3(1 - 2x)^3]$ cm^3/mol, of Aqueous Mixtures at 298.15 K, Where x Is the Mole Fraction of the Cosolvent

Cosolvent	Ref.	v_0	v_1	v_2	v_3
Methanol	a	−4.08	−0.43	0.69	1.20
Ethanol	a	−4.37	−1.70	−0.74	1.64
2-Propanol	b	−3.59	−2.28	−2.49	
2-Methyl-2-propanol	c	−2.18	−2.08	−4.78	
1,2-Ethanediol	d	−1.46	−0.56	0.22	
Glycerol	e	−1.35	−0.25	−0.84	−0.97
Tetrahydrofuran	f	−3.09	−2.35	−1.25	
Dioxane	g	−2.45	−1.90	−0.88	−0.54
Acetone	h	−5.79	−2.24	−0.34	
Formic acid	i	−1.50	−0.70	−0.83	
Acetic acid	i	−4.60	0.50	−1.84	
Triethylamine	j	−10.15	−0.50	−2.54	−3.75
Pyridine	k	−2.93	−0.97	0.85	
Acetonitrile	l	−1.99	−2.04	−1.32	
Formamide	m	−0.51	−0.51	−0.31	
N-Methylformamide	n	−2.13	−1.05	0.14	
N, N-Dimethylformamide	o	−4.70	−1.22	1.71	
N-Methylpyrrolidin-2-one	o	−4.59	−1.93	0.52	
Hexamethyl phosphoric triamide	p	−5.43	−4.65	−4.43	
Dimethylsulfoxide	m	−3.90	−2.09	1.20	2.04

[a]GC Benson, O Kiyohara. J Solution Chem 9:791, 1980.
[b]M Sakurai. J Solution Chem 17:267, 1988.
[c]JF Alary, MA Simard, J Dumont, C Jolicoeur. J Solution Chem 11:755, 1982.
[d]J-Y Huot, E Battistel, R Lumry, G Villeneuve, J-F Lavallee, A Anusiem, C Jolicoeur. J Solution Chem 17:601, 1988.
[e]JM Saleh, M Khalil, NA Hikmat. J Iraqi Chem Soc 11:89, 1986; ECH To, JV Davies, M Tucker, P Westh, C Trandum, KSH Suh, Y Koga. J Solution Chem 28:1137, 1999 (up to $x = 0.4$).
[f]O Kiyohara, GC Benson. Can J Chem 55:1354, 1977.
[g]MBR Paz, AH Buep, M Baron. An Asoc Quim Argent 76:69, 1988.
[h]K Noda, M Ohashi, K Ishida. J Chem Eng Data 27:326, 1982.
[i]A Apelblat, E Manzurola. Fluid Phase Equil 32:163, 1987.
[j]TM Letcher, W Spiteri. J Chem Thermodyn 11:905, 1979.
[k]J-I Abe, K Nakanishi, H Touhara. J Chem Thermodyn 10:483, 1978.
[l]AJ Easteal, LA Woolf. J Chem Thermodyn 20:701, 1988.
[m]C deVisser, WJM Heuvelsland, LA Dunn, G Somsen. J Chem Soc Faraday Trans 1 74:1159, 1978.
[n]C deVisser, P Pel, G Somsen. J Solution Chem 6:571, 1977.
[o]MI Davis. Thermochim Acta 120:299, 1987.
[p]M Castagnolo, A Inglese, G Petrella, A Sacco. Thermochim Acta 44:67, 1981.

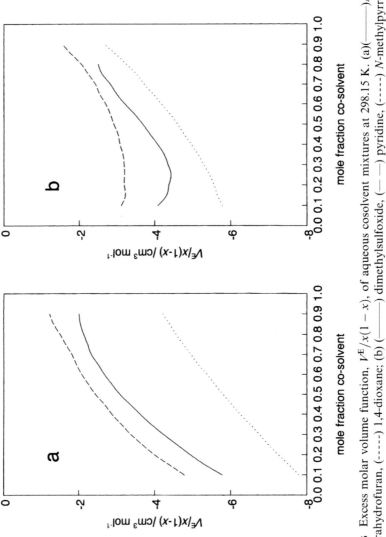

FIGURE 2.5 Excess molar volume function, $V^E/x(1-x)$, of aqueous cosolvent mixtures at 298.15 K. (a)(———)Acetone, (———)tetrahydrofuran, (-----) 1,4-dioxane; (b) (———) dimethylsulfoxide, (— —) pyridine, (-----) N-methylpyrrolidin-2-one.

TABLE 2.17A Excess Volume, $V^E/cm^3\ mol^{-1}$, of Mixtures of Benzene with Cosolvents, $V^E = x(1-x)\sum v_i(1-2x)^i$, Where x Is the Mole Fraction of the Cosolvent

Cosolvent	Ref.	T/K	v_0	v_1	v_2	v_3	v_4
Hexane	a	298.15	1.651	0.056	0.066		
Cyclohexane	a	298.15	2.620	-1.117	0.044		
Toluene	a	298.15	0.176	-0.093			
Chloroform	a	293.15	0.608	0.160	0.102		
Tetrachloromethane	a	298.15	0.017	-0.019	0.031		
Methanol	a	293.15	-0.030	-0.253	0.029	0.094	-0.028
Ethanol	b	298.15	0.087	0.603	0.202	0.302	
Ethyl ether							
Tetrahydrofuran	a	298.15	-1.014	-0.123	0.064		
Dioxane	a	298.15	-0.261	-0.247	-0.058	0.022	0.018
Acetone	a	298.15	-1.289	-0.026	0.089		
Ethyl acetate	a	298.15	0.469	0.511	-0.657	0.121	
Pyridine	a	298.15	-0.799	-0.144			
Nitromethane	c	298.15	0.908	-0.643	-0.005		
Nitrobenzene	a	298.15	-0.576	-0.301			
Acetonitrile	a	303.15	-0.307	-0.910			
Dimethylformamide	a	303.15	-1.09	-1.84			
Dimethylsulfoxide	a	303.15	-1.341	0.725	-0.302		

Refs. can be found on pgs. 81 and 82.

This is true not only for the V^E data but also for other thermodynamic quantities described by expressions similar to Eq. (2.35).

In certain cases the *partial molar volumes* of the components of the mixture, v_A and v_B, i.e., the actual fractions of the total volume of the mixture ascribable to them,

$$x_A v_A + x_B v_B = V \tag{2.38}$$

are required. These are obtained from the general thermodynamic expressions for partial molar quantities in binary mixtures, written in terms of the volumes:

$$v_A = V - x_B dV/dx_B = V_A + V^E - x_B dV^E/dx_B \tag{2.39a}$$

$$v_B = V + x_A dV/dx_B = V_B + V^E + x_A dV^E/dx_B \tag{2.39b}$$

ensuring that Eq. (2.38) is obeyed. The partial molar volume of the cosolvent at infinite dilution in water, v_B^∞, was shown by Gianni and Lepori

TABLE 2.17B Excess Volume, $V^E/cm^3 mol^{-1}$, of Mixtures of Tetrachloromethane with Cosolvents, $V^E = x(1-x) \sum v_i(1-2x)^i$, Where x Is the Mole Fraction of the Cosolvent

Cosolvent	Ref.	T/K	v_0	v_1	v_2	v_3	v_4
Hexane	a	298.15	0.178	0.003	0.031		
Cyclohexane	a	298.15	0.658	-0.007	0.028		
Benzene	a	298.15	0.017	0.019	0.031		
Toluene							
Chloroform	d	298.15	0.701	0.024	0.106		
Methanol	a	293.15	-0.156	0.233	0.129	0.201	
Ethanol							
Ethyl ether	S-73	298.15	-2.846	0.463	0.076		
Tetrahydrofuran	a	303.15	-2.393	0.551	-0.156		
Dioxane	a	303.15	-1.36	-0.705	0.549		
Acetone	a	298.15	-0.443	0.002	0.298		
Ethyl acetate	a	298.15	0.20	0.05	0.2		
Pyridine							
Nitromethane	c	298.15	0.677	0.757	0.503	1.211	1.207
Nitrobenzene							
Acetonitrile	a	298.15	-0.530	-0.050	0.367		
Dimethylformamide	a	303.15	-5.05	-1.10	-3.26		
Dimethylsulfoxide	a	298.15	-2.437	0.137	0.267	-0.110	-0.089

Refs. can be found on pgs. 81 and 82.

[20] to be calculable from group contributions. As an example of the application of Eq. (2.39), the excess partial molar volumes of water (A) and ethanol (B) in aqueous ethanol, $v_A - V_A$ and $v_B - V_B$, are shown for 298.15 K in Fig. 2.6, in addition to the excess molar volume. Note that the curve for ethanol has a minimum that would be much more pronounced if the plot were made at a lower temperature [21].

The isobaric expansivity, α_P, is the measurable relative increase of the volume or decrease of the density on unit increase of the temperature at constant pressure. Thus, $\alpha_P = (\partial \ln V/\partial T)_P = -(\partial \ln d/\partial T)_P$, and is of the order of $10^{-3} K^{-1}$. For mixtures of water A with cosolvent B it can be written as the sum of the mole fraction–weighted contributions of the neat components and a polynomial of the form of the Redlich–Kister expression:

$$\alpha_P/(10^{-3} K^{-1}) = (1-x_B)\alpha_{PA} + x_B\alpha_{PB} + x_B(1-x_B)[\alpha_0 + \alpha_1(1-2x_B)$$
$$+ \alpha_2(1-2x_B)^2 + \alpha_3(1-2x_B)^3 + \cdots] \qquad (2.40)$$

TABLE 2.17C Excess Volume, $V^E/cm^3\ mol^{-1}$, of Mixtures of Methanol with Cosolvents, $V^E = x(1-x)\sum v_i(1-2x)^i$, Where x Is the Mole Fraction of the Cosolvent

Cosolvent	Ref.	T/K	v_0	v_1	v_2	v_3	v_4
Hexane*		298.15		Miscibility gap			
Cyclohexane		298.15		Miscibility gap			
Benzene	a	293.15	−0.030	0.253	0.029	−0.094	−0.028
Toluene	e	298.15	−1.507	2.452	−1.337	3.066	
Chloroform	f	298.15	−0.541	−0.273	−0.056		
Tetrachloromethane	a	293.15	−0.156	−0.233	0.129	−0.201	
Ethanol	a	298.15	0.034	0.009	0.010		
Ethyl ether	g	298.15	−3.153	0.148	−0.642		
Tetrahydrofuran	a	298.15	−0.601	−0.168	−0.003	0.063	
Dioxane	a	303.15	−1.516	−0.498	0.123		
Acetone	a	298.15	−1.48	0.225	−0.820		
Ethyl acetate	h	293.15	−0.367	0.010	0.014		
Pyridine	a	298.15	−1.955	−0.841	−0.101	0.084	
Nitromethane	a	298.15	−0.723	−0.441	−0.351		
Nitrobenzene	i	298.15	−3.378	−6.789	14.837	−8.223	
Acetonitrile	a	298.15	−0.636	−0.375	0.017	−0.199	−0.424
Dimethylformamide	a	303.15	−1.701	−0.804	0.201		
Dimethylsulfoxide	a	298.15	−2.14	−0.192	0.371	1.12	

*Outside the range $0.211 \le x \le 0.736$; inside this range phase separation takes place.
Refs. can be found on pgs. 81 and 82.

where $\alpha_{PA} = 0.257 \times 10^{-3}\ K^{-1}$ for water at 298.15 K. The values of α_{PB} and the coefficients α_{P_j} for common aqueous cosolvent mixtures at 298.15 K are shown in Table 2.18.

The adiabatic compressibilities, $\kappa_S = -(\partial \ln V/\partial P)_S$, of mixtures of solvent A with cosolvents B are obtained from measurements of the speed of ultrasound, u, (of the order of 1 km sec^{-1}) in and the density, d, of the mixtures: $\kappa_S = 1/du^2$. The resulting values are of the order of 1 GPa^{-1}. They can again be written in terms of a Redlich–Kister-type expression:

$$\kappa_S/GPa^{-1} = (1-x_B)\kappa_{SA} + x_B\kappa_{SB} + x_B(1-x_B)[\kappa_0 + \kappa_1(1-2x_B)$$
$$+ \kappa_2(1-2x_B)^2 + \kappa_3(1-2x_B)^3 + \cdots] \qquad (2.41)$$

The values of κ_{SB}, and the coefficients κ_i for common aqueous cosolvent mixtures at 298.15 K are shown in Table 2.19, where $\kappa_{SA} = 0.448\ GPa^{-1}$

TABLE 2.17D Excess Volume, $V^E/cm^3 \ mol^{-1}$, of Mixtures of Ethanol with Cosolvents, $V^E = x(1-x)\sum v_i(1-2x)^i$, Where x Is the Mole Fraction of the Cosolvent

Cosolvent	Ref.	T/K	v_0	v_1	v_2	v_3	v_4
Hexane	a	298.15	1.71	−0.649	1.25	−1.28	
Cyclohexane	a	298.15	2.260	−0.090	1.036		
Benzene	b	298.15	0.087	−0.603	0.202	−0.302	
Toluene	b	298.15	−0.249	−0.671	0.104	−0.277	
Tetrachloromethane							
Chloroform							
Methanol	a	298.15	0.034	−0.009	0.010		
Ethyl ether	g	298.15	−2.648	0.151	0.660		
Tetrahydrofuran	j	298.15	−0.084	−0.254			
Dioxane	k	298.15	−1.807	−0.035	1.088	0.121	−1.560
Acetone	h	293.15	−0.308	−0.144	−0.068		
Ethyl acetate	l	298.15	0.631	−0.141	−0.057	0.345	
Pyridine	m	298.15	−1.49	−0.82			
Nitromethane	a	298.15	−0.579	−0.173	−0.566	−0.365	
Nitrobenzene	i	298.15	−2.685	−4.098	2.644	2.598	
Acetonitrile	a	298.15	−0.114	−0.644	−0.130		
Dimethylformamide	a	303.15	−0.972	−0.426	−0.438	1.025	
Dimethylsulfoxide	n	298.15	−1.457	−0.051	−0.591	0.550	

Refs. can be found on pgs. 81 and 82.

for water at 298.15 K. The isothermal compressibilities, $\kappa_T = -(\partial \ln V / \partial P)_T = (\partial \ln d/\partial P)_T$, are generally some 10% larger. Although they have been obtained from density measurements at high pressures (of the order of 100 MPa and more) for many pure solvents (see Ref. 2), they have not been widely reported in the literature for mixtures of solvents. They can be calculated from

$$\kappa_T = \kappa_S + TV\alpha_P^2/C_P \tag{2.42}$$

This calculation requires values for the isobaric expansivity, α_P, the molar volume, V, and isobaric heat capacity, C_P, that can be calculated from the data in Tables 2.18, 2.16 and 2.15, respectively.

The surface properties of solvent mixtures are described by the surface tension, σ. This quantity is directly measurable and describes the force per unit length or the energy per unit area of the surface of the liquid mixture relative to vacuum, in principle, or relative to air saturated with

TABLE 2.17E Excess Volume, $V^E/\text{cm}^3 \text{ mol}^{-1}$, of Mixtures of Acetone with Cosolvents, $V^E = x(1 - x) \sum v_i (1 - 2x)^i$, Where x Is the Mole Fraction of the Cosolvent

Cosolvent	Ref.	T/K	v_0	v_1	v_2	v_3	v_4
Hexane	o	298.15	4.253	−0.517	0.521		
Cyclohexane	a	298.15	4.482	−0.054	0.741	−0.430	
Benzene	a	298.15	−1.289	0.026	0.089		
Toluene	p	298.15	−0.646	0.043	−0.010		
Tetrachloromethane	a	298.15	−0.443	−0.002	0.298		
Chloroform	q	298.15	−0.424	−1.011	0.058	0.259	
Methanol	a	298.15	−1.48	−0.225	−0.820		
Ethanol	h	293.15	−0.308	0.144	−0.068		
Ethyl ether							
Tetrahydrofuran							
Dioxane							
Ethyl acetate							
Pyridine							
Nitromethane							
Nitrobenzene							
Acetonitrile	r	298.15	−0.485	−0.037			
Dimethylformamide							
Dimethylsulfoxide	S-80	298.15	−2.044	−0.456	−0.120		

Refs. can be found on pgs. 81 and 82.

the vapors, in practice. It is formally defined as $\sigma = (\partial G/\partial A)_{T,P,ni}$, where A is the surface area of the liquid mixture. For aqueous mixtures at 298.15 K, the dependence of σ on the composition is conveniently described as

$$\sigma/\text{mN m}^{-1} = (1 - x_B)\sigma_A + x_B\sigma_B + \sigma_0/[1 - \sigma_1(1 - x_B)] \tag{2.43}$$

The surface tension of water at 298.15 K is $\sigma_A = 71.8 \text{ mN m}^{-1}$. Because the surface tension decreases steeply when a cosolvent is added to water, a Redlich–Kister-type expression analogous to Eq. (2.41), with σ_j replacing κ_j, is generally less suitable for aqueous mixtures. Expressions arising from the concepts of the competitive solvation model, Section 4.4, in particular the two-step model, have been found by Khossravi and Connors [21a] to be useful. According to this model,

$$\sigma = \sigma_A - (\sigma_A - \sigma_B)[(\beta_1 x_A x_B + 2\beta_2 x_A^2)/2(x_A^2 + \beta_1 x_A x_B + \beta_2 x_A^2)] \tag{2.44}$$

TABLE 2.17F   Excess Volume, $V^E/cm^3\ mol^{-1}$, of Mixtures of Acetonitrile with Cosolvents, $V^E = x(1-x)\sum v_i(1-2x)^i$, Where x Is the Mole Fraction of the Cosolvent

Cosolvent	Ref.	T/K	v_0	v_1	v_2	v_3	v_4
Hexane		298.15	Miscibility gap				
Cyclohexane		298.15	Miscibility gap				
Benzene	r	298.15	−0.199	0.782	−0.038		
Toluene							
Tetrachloromethane	a	298.15	−0.530	0.050	0.367		
Chloroform	r	298.15	0.730	0.197	−0.358		
Methanol	a	298.15	−0.636	0.375	0.017	0.199	−0.424
Ethanol	a	298.15	−0.114	0.644	−0.130		
Ethyl ether	s	298.15	−3.290	0.260	−0.406	0.268	
Tetrahydrofuran	s	298.15	−0.600	0.296	−0.107	−0.136	
Dioxane	r	298.15	−1.053	−0.194			
Acetone	r	298.15	−0.485	0.037			
Ethyl acetate							
Pyridine							
Nitromethane	t	298.15	−1.483	−0.044	0.097		
Nitrobenzene	u	293.16	−1.712	−0.825	−0.250	0.142	
Dimethylformamide	r	298.15	−0.537	−0.135			
Dimethylsulfoxide	r	298.15	−0.692	−0.271			

Refs. can be found on pgs. 81 and 82.

where the β's are equilibrium constants for replacement of water molecules (A) by one and two cosolvent molecules (B), describes aqueous systems well. Nevertheless, a Redlich–Kister-type expression can be applied to nonaqueous ones, for which the difference $\sigma_A - \sigma_B$ is much smaller than for aqueous systems. The values of the coefficients σ_j and of σ_B for those of the aqueous mixtures of the common cosolvents where the surface tension at 298.15 K is known are shown in Table 2.20. For the non-aqueous solvent mixtures a different way of reporting the data is preferred. They are reported for mixtures of each of the seven solvents A at definite mole fractions x_B: 0.00, 0.15, 0.35, 0.50, 0.65, 0.85, and 1.00 of the 18 solvents B (where known) in Table 2.21. Note again that the value of σ or any other property) at a certain x_B (say, 0.85) of solvent B in the table of mixtures involving solvent A, must be the same as the value of σ at $x_A = 1 - x_B$ (i.e., 0.15) of solvent A in the table involving various mixtures of solvent B.

The composition of the surface layer in a binary mixture is generally different from that of the bulk, because one of the components may be

TABLE 2.17G   Excess Volume, V^E/cm^3mol^{-1}, of Mixtures of Nitromethane with Cosolvents, $V^E = x(1-x)\sum v_i(1-2x)^i$, Where x Is the Mole Fraction of the Cosolvent

Cosolvent	Ref.	T/K	v_0	v_1	v_2	v_3	v_4
Hexane	c	298.15	Miscibility gap: $0.026 \le x \le 0.964$				
Cyclohexane	c	298.15	Miscibility gap: $0.037 \le x \le 0.969$				
Benzene	c	298.15	0.908	0.610	−0.005	0.100	
Toluene							
Tetrachloromethane	c	298.15	0.663	−0.258	0.324	−0.320	
Chloroform							
Methanol	a	298.15	−0.723	0.441	−0.351		
Ethanol	a	298.15	−0.579	0.173	−0.566	0.365	
Ethyl ether							
Tetrahydrofuran							
Dioxane	a	303.15	−0.092	0.022	0.044		
Acetone							
Ethyl acetate	v	298.15	−0.713	−0.145	−0.218	0.148	
Pyridine							
Nitrobenzene							
Acetonitrile	t	298.15	−0.483	0.044	0.097		
Dimethylformamide							
Dimethylsulfoxide	w	318.15	0.513	0.187	0.335	−0.909	

[a] Y Handa, GC Benson. Fluid Phase Equil 3:185,1979.
[b] R Tanaka, S Toyama J Chem Eng Data 42:871,1997.
[c] KN Marsh. J Chem Thermodyn 17:29,1985.
[d] M Artal, JM Embid, I Velasco, S Otin. J Chem Thermodyn 23:1131, 1991.
[e] RK Wanchoo, J Narayan. Phys Chem Liq 25:15,1992.
[f] J Winnick, J Kong. Ind Eng Chem Fundam 13:292, 1974.
[g] J Canosa, A Rodriguez, J Tojo. Fluid Phase Equil 156:57, 1999.
[h] A Qin, DE Hoffman, P Munk. Coll Czech Chem Commun 58:2625, 1993.
[i] PS Nikam, MC Jadhav, M Hasan. J Chem Eng Data 40:931, 1995.
[j] L Lepori, E Matteoli. Fluid Phase Equil 145:69, 1998.
[k] MSh Ramadan, AM Hafez, AA El-Zyadi. Phys Chem (Peshawar, Pakistan) 10:55, 1991.
[l] JH Hu, K Tamura, S Murakami. Fluid Phase Equil 134:239, 1997.
[m] TJV Findlay, JL Copp.Trans Faraday Soc 65:1463, 1969.
[n] PS Nikam, MC Jadhav, M Hasan. J Chem Eng Data 41: 1028, 1996.
[o] G Marino, MM Pineiro, M Iglesias, B Orge, J Tojo. J Chem Eng Data 46:728 2001.
[p] J Nath, AP Dixit. J Chem Eng Data 28:190, 1983.
[q] Y Akamatsu, H Ogawa, S Murakami. Thermochim. Acta 113:141, 1987.
[r] J-PE Grolier, G Roux-Desgranges, M Berkane, E Wilhelm. J Chem Thermodyn 23:421, 1991.
[s] TM Letcher, U Domanskaya. J Chem Thermodyn 26:113, 1994.

TABLE 2.17   Footnotes continued

[t]A D'Aprano, A Capalbi, M Iammarino, V Mauro, A Princi, B Sesta. J Solution Chem 24:227, 1995.
[u]L Jannelli, M Pansini, A Lopez. Fluid Phase Equil 15:219, 1983.
[v]S-L Lee, C-H Tu. J Chem Eng Data 44:108, 1999.
[w]LS Manjeshwar, TM Aminabhavi. J Chem Eng Data 32:409, 1987.
[s-nn]International Data Series, Selected Data Mixtures, Ser A, 19nn.

preferentially concentrated in (adsorbed at) the surface in order to reduce the total Gibbs energy of the system. If the vapor phase can be considered ideal and the amount of component B in it is negligible, then its (excess mole fraction) surface concentration, Γ_B, in the liquid mixture is

$$\Gamma_B = -(1/RT)[\partial\sigma/\ln x_B]_T \tag{2.45}$$

The surface tension of aqueous mixtures with cosolvents, σ, decreases with increasing contents of the cosolvent and $\ln x_B$ becomes less negative. Therefore, the derivative in square brackets in Eq. (2.45) is negative and the surface concentration of the cosolvent is positive, that of water, $\Gamma_A = -\Gamma_B$, being negative. The physical meaning of this is that a cosolvent that decreases the surface tension of water (as practically all cosolvents do) concentrates at the surface at the expense of the water. The excess surface concentration of a component times the surface area A is the amount of this component in the hypothetical surface "phase" beyond the amount present in the bulk. However, Γ_i cannot be compared with x_i because the thickness (or volume) of the surface "phase" is ill defined.

The critical temperature and pressure of a solvent mixture can be estimated on returning to the PVT, i.e., the pressure–volume–temperature, properties of the mixture. A generally useful assumption in this connection is that of the *corresponding states*. This assumes that the PVT properties of the mixture are the same as that of a pure component with critical temperature T_C and pressure P_C equaling the *pseudocritical* temperature T_{Cp} and pressure P_{Cp} of the mixture. The pseudocritical temperature of a binary mixture is approximated by the mole fraction weighted average:

$$T_{Cp} = x_A T_{CA} + x_B T_{CB} \tag{2.46}$$

and the pseudocritical pressure is approximated as

$$P_{Cp} = RT_{Cp}[x_A Z_{CA} + x_{BA} Z_{CB}]/[x_A V_{CA} + x_B V_{CB}] \tag{2.47}$$

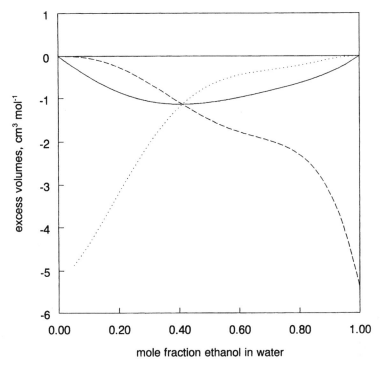

FIGURE 2.6 Excess molar volume of aqueous ethanol (———) and excess partial molar volume of the water (— —) and of the ethanol (-----) at 298.15 K.

where Z_C is the critical compressibility factor $P_C V_C / R T_C$. More sophisticated values have been estimated for mixtures of the solvents on our list according to Escobedo and Mansoori [22], with the T_{Cp} results shown in Table 2.22. The true critical temperatures of binary mixtures T_{Ct} are given to a good approximation by the critical temperatures of the components weighted by the critical volume fraction, as

$$T_{Ct} = [x_A V_{CA} T_{CA} + x_B V_{CB} T_{CB}] / [x_A V_{CA} + x_B V_{CB}] \qquad (2.48)$$

Critical temperatures of a few solvent mixtures (aqueous dioxane, THF, methanol, ethanol, and acetone and mixtures of acetone with ethanol and of benzene with toluene) as functions of the composition were reported by Marshall and Jones [23].

TABLE 2.18 Coefficients α_j of the Isobaric Expansivity, $1000(\alpha_P/K^{-1}) = 0.256(1-x) + 1000\alpha_{cosolvent}x + x(1-x)[\alpha_0 + \alpha_1(1-2x) + \alpha v_2(1-2x)^2 + \alpha_3(1-2x)^3]$, of Aqueous Mixtures at 298.15 K, Where x is the Mole Fraction of the Cosolvent

Cosolvent	Ref.	$1000\alpha_{cosolvent}$	α_0	α_1	α_2	α_3
Methanol	a	1.198	0.868	0.554	−0.031	
Ethanol	b	1.096	1.410	1.019	0.0336	
2-Propanol	c	1.083	0.987	0.0063	0.618	
2-Methyl-2-propanol	d	1.20	1.19	2.31	1.49	
1,2-Ethanediol	e	0.838	0.316	1.118	−0.217	
Glycerol	f	0.290	0.282			
Tetrahydrofuran	g	1.290	0.70	1.71		
Dioxane	h	1.093	1.088	1.036	1.118	
Acetone	i	1.40	1.48	0.84	0.91	
Formic acid	j	1.031	0.932	0.983	0.840	
Acetic acid	j	1.075	1.034	1.001	1.107	
Triethylamine (at 287.15 K)	k	0.94	0.989	−0.195	5.909	12.646
Pyridine						
Acetonitrile	l	1.390	1.269	0.910	1.018	
Formamide						
N-Methylformamide	m	0.876	0.902	0.773	0.468	
N,N-Dimethylformamide	n	1.036	1.380	0.894		
N-Methylpyrrolidin-2-one						
Hexamethyl phosphoric triamide	o	0.830	1.010	2.000	3.341	
Dimethylsulfoxide						

[a]GC Benson, O Kiyohara. J. Solution Chem 9:791, 1980.
[b]G D'Arrigo, A Paparelli. J Chem Phys 88:405, 1988.
[c]M Sakurai. J Solution Chem 17:267, 1988.
[d]JF Alary, MA Simard, J Dumont, C Jolicoeur. J Solution Chem 11:755, 1982.
[e]M Tsierkezos, IE Molinou. J Chem Eng Data 43:989, 1998.
[f]NM Murthy, SV Subrahmanyam. Indian J Pure Appl Phys 15:485, 1977.
[g]A Wencel, K Czerrepko. Pol J Chem 65:1809, 1991.
[h]MBR Paz, AH Buep, M Baron. An Asoc Quim Argent 76:69, 1988.
[i]GI Egorov, EL Gruznov, AM Kolker. Russ J Phys Chem 70:197, 1996.
[j]A Apelblat, E Manzurola. Fluid Phase Equil 32:163, 1987.
[k]TM Letcher, W Spiteri, J Chem Thermodyn 11:905, 1979.
[l]AJ Easteal, LA Woolf. J Chem Thermodyn 20:701, 1988.
[m]C deVisser, P Pel, G Somsen. J Solution Chem 6:571, 1977.
[n]AA Ennan, AN Chobomarev, VA Anike'ev, ME Kornelli, EI Yur'eva, BM Kats. Zh Priklad Khim (Leningrad) 45:622, 1972.
[o]MB Kulikov, AM Kolker, AG Krestov. Zh Priklad Khim (Leningrad) 63:852, 1990.

TABLE 2.19 Coefficients κ_j of the Adiabatic Compressibility, $\kappa_S = 0.448(1 - x) + \kappa_{cosolvent}x + x(1 - x)[\kappa_0 + \kappa_1(1 - 2x) + \kappa_2(1 - 2x)^2 + \kappa_3(1 - 2x)^3]$ GPa^{-1}, of Aqueous Cosolvent Mixtures at 298.15 K, Where x is the Mole Fraction of the Cosolvent

Cosolvent	Ref.	$\kappa_{cosolvent}$	κ_0	κ_1	κ_2	κ_3
Methanol	e	1.049	−0.655	−0.228	−0.102	
Ethanol	c	0.948	−0.171	−0.166	−0.495	−0.963
2-Propanol	b	0.992	0.129	0.085	−0.791	−1.065
2-Methyl-2-propanol	g	0.920	1.454	−13.34	33.68	−25.11
1,2-Ethanediol	a	0.351	−0.309	−0.139	−0.300	0.159
Glycerol	n	(0.240)	−0.281	0.253	1.043	
Glycerol*	d	0.246	−0.350	−0.292	−0.360	−0.305
Tetrahydrofuran	a	0.682	0.204	−0.073	−0.512	−0.641
Dioxane	a	0.523	−0.045	−0.388	−0.312	
Acetone*	k	1.257	−0.047	−0.471	−0.616	
Formic acid*	j	0.656	−0.152	−0.136	−0.148	
Acetic acid*	j	0.930	−0.272	−0.090	−0.428	
Triethylamine						
Pyridine	m	0.507	-0.027_7	0.003_5	-0.014_6	
Acetonitrile	a	0.782	0.154	−0.086	−0.382	−0.694
Formamide*	f	0.399	−0.173	−0.128		
N-Methylformamide	i	0.489	−0.189	−0.243	−0.239	−0.228
N,N-Dimethylformamide	a	0.503	−0.321	−0.380	−0.153	
	i	0.498	−0.287	−0.305	0.359	−0.335
N-Methylpyrrolidin-2-one	l	0.408	−0.650	−0.100	−0.747	−0.880
Hexamethyl posphoric triamide	h	0.553	−0.096	0.599	−0.461	−2.836
	o	0.524	−0.065	−0.257	−0.436	
Dimethylsulfoxide	a	0.411	−0.408	−0.330	−0.321	

*For these cosolvents, the entries pertain to the isothermal compressibility, κ_T.

[a] TM Aminabhavi, B Gopalakrishna. J Chem Eng Data 40:856, 1995.
[b] G Douheret, MM Holczer, R Peyrelier. J Chem Eng Data 39:868, 1994.
[c] G D'Arrigo, A Paparelli. J Chem Phys 88:405, 1988 (mean of 20 and 30°C data); G Omori. J Chem Phys 89:4325, 1988.
[d] Y Miyamoto, M Takemoto, M Hosokawa, Y Uosaki, T Moriyoshi. J Chem Thermodyn 22:1007, 1990.
[e] K Jerie, A Baranowski, B Rozenfeld, S Ernst, B Jezowska-Trzebiatowska, J Glinski. Acta Phys Pol A66:167, 1984.
[f] M Oguni, CA Angell. J Chem Phys 78:7334, 1983.
[g] CM Sehgal, BR Porter, JF Greenleaf. J Acoust Soc Am 79:566, 1986 (up to $x \leq 0.3$).
[h] AK Lyashchenko, VS Goncharov, T Iijima, H Uedaria, J Komiyama. Bull Chem Soc Jpn 53:1888, 1980 (up to $x \leq 0.4$).
[i] F Kawaizumi, M Ohno, Y Miyahara. Bull Chem Soc Jpn 50:2229, 1977.

TABLE 2.19 Footnotes continued

[j]J Korpela. Acta Chem Scand 25:2852, 1971.
[k]GI Egorov, EL Gruznov, AM Kolker. Zh Fiz Khim 70:216, 1996.
[l]A Pal, YP Singh, W Singh. Indian J Chem 33A:1083, 1994.
[m]VK Sharma, J Singh. Indian J Chem 37A:59, 1998.
[n]EL Ravi Meher, KC Reddy. Indian J Pure Appl Phys 25:22, 1987 (the value for neat glycerol is approximate).
[o]DK Jha, BL Jha. Acustica 75:279, 1992 (from $x \geq 0.2$ upward).

2.4 CHEMICAL PROPERTIES

The dissolution of a solute in a solvent or in solvent mixtures can be said to involve the work required for the formation of a cavity sufficiently large to accommodate its molecules or ions (see Section 1.2). The structuredness [2,24] of the solvent mixture is a feature that is relevant to this amount of work. There are several measures of the structuredness of solvents, which describe properties such as "order," "stiffness," and "openness." The question of whether these can be applied to solvent mixtures too has not been answered satisfactorily so far. It must, in fact, be conceded that none of these measures has been quantitatively applied to solvent mixtures, but some are readily calculated from the properties of the neat solvents and the composition of the mixture.

The *stiffness* of a solvent is measured by its cohesive energy density, and this concept may also be transferred to solvent mixtures. The cohesive energy, E^C, is the energy that has to be invested in a liquid in order to separate its molecules to an infinite distance apart, i.e., to vaporize it to an ideal vapor. This configurational energy is, thus, the internal energy of the vapor molecules less the sum of the internal energy of the liquid molecules and their energy of interaction. Per mole of the solvent mixture this difference is $E^C = \Delta_V U = \Delta_V H - RT$, translating the enthalpy change on vaporization into the energy change by allowing for the translational energy of the ideal vapor. For a binary solvent mixture this quantity is

$$E^C = x_A \Delta_V H_A + x_B \Delta_V H_B - RT - H^E \quad (2.49)$$

The cohesive energy density of a neat solvent is E^C / V and is the square of its *solubility parameter* δ. Hence, a solubility parameter can be defined for the mixed solvent:

$$\delta_{AB}^2 = [x_A \Delta_V H_A + x_B \Delta_V H_B - RT - H^E] / [x_A V_A + x_B V_B + V^E] \quad (2.50)$$

TABLE 2.20 Coefficients σ_j of the Surface Tension, $\sigma = 71.8(1 - x) + \sigma_{cosolvent}x + x(1 - x)f(x)\,mN\cdot m^{-1}$, Where Generally $f(x) = \sigma_0/[1 - \sigma_1(1 - x)]$, Except in the Cases Marked by *, Where $f(x) = \sigma_0 + \sigma_1(1 - 2x) + \sigma_2(1 - 2x)^2 + \sigma_3(1 - 2x)^3$, of Aqueous Cosolvent Mixtures at 298.15 K, x Being the Mole Fraction of the Cosolvent

Cosolvent	Ref.	$\sigma_{cosolvent}$	σ_0	σ_1	σ_2	σ_3
Methanol	c*	22.4	−68	−38	−57	−106
Ethanol	e*	21.8	−97	−196	746	−1155
2-Propanol	c	20.8	−49.1	0.983		
2-Methyl-2-propanol	c	19.9	−50.6	0.992		
1,2-Ethanediol	a*	48.9	−28.5	−12.1	−27.8	−54.2
Glycerol	f	63.3	−3.1	0.771		
Tetrahydrofuran	c	26.9	−43.1	0.983		
Dioxane	g	32.8	−35.05	0.9637		
Acetone	d	23.1	−40.39	0.981		
Formic acid	b	37.0	−26.36	0.846		
Acetic acid	b	26.9	−35.5	0.956		
Triethylamine		20.1				
Pyridine	g	37.0	−23.7	0.993		
Acetonitrile	c	28.3	−41.6	0.953		
Formamide	f	58.2	−11.3	0.754		
N-Methylformamide		39.5				
N,N-Dimethylformamide	g*	35.2	−49.18	−17.01	−60.34	−130.91
N-Methylpyrrolidin-2-one		40.7				
Hexamethyl phosphoric triamide		33.8				
Dimethylsulfoxide	c*	43.0	−37	4	35	−135

[a]NG Tsierkezos, IE Molinou. J Chem Eng Data 43:989, 1998.
[b]E Alvarez, G Vazquez, M Sanchez-Vilas, B Sanjurjo, J M Navaza. J Chem Eng Data 42:957, 1997.
[c]WJ Cheong, PW Carr. J Liq Chromatogr 10:561, 1987.
[d]AI Toryanik, VG Pogrebnyak. Zh Strukt Khim 17:536, 1976.
[e]FW Seemann. PTB-Mitteilungen 91:95, 1981.
[f]H-D. Dörfler. Colloid Polym Sci 257:882, 1979.
[g]DR Lide, ed. Handbook of Chemistry and Physics. 82nd ed. Baton Rouge: CRC Press, 2001–2002, p. 6–150.

TABLE 2.21A   Surface Tension σ at 298.15 K of Mixtures of Benzene ($\sigma = 28.2\,\mathrm{mN\,m^{-1}}$) with Cosolvents at the Stated Mole Fractions of the Cosolvent

Cosolvent	Ref.	0.15	0.35	0.50	0.65	0.85	1.00
Hexane	a	24.8	22.0	20.7	19.9	18.9	17.94
Cyclohexane	b	26.9	25.8	25.2	24.8	24.5	24.30
Toluene	T	27.81	27.91	28.05	28.18	28.45	28.93
Chloroform	T	27.92	27.6	27.37	27.14	26.85	26.65
Tetrachloromethane	T	28.06	27.79	27.52	27.20	26.66	26.19
Methanol	T	27.60	26.63	25.79	24.87	23.50	22.36
Ethanol	T	26.93	25.80	24.96	24.12	22.99	22.15
Ethyl ether	T	25.18	22.27	20.28	18.43	16.62	16.35
Tetrahydrofuran							
Dioxane							
Acetone	T	27.51	26.51	25.71	24.87	23.67	22.73
Ethyl acetate	c	27.3	26.4	25.8	25.1	24.2	23.6
Pyridine							
Nitromethane	d	28.8	29.0	29.7	30.9	33.2	35.5
Nitrobenzene	T	29.40	31.68	33.69	35.96	39.39	42.25
Acetonitrile							
Dimethylformamide							
Dimethylsulfoxide							

Refs. can be found on pg. 94.

that might be a useful quantity. Ordinarily, the excess quantities H^E and V^E are small with respect to the molar enthalpies of vaporization and the molar volumes. For a neat unstructured solvent the cohesive energy density E^C/V is within 50 J cm^{-3} of the internal pressure, $P_{in} = T\alpha_P/\kappa_T - P$ as found by Marcus [24], but for structured solvents this difference is considerably larger, $E^C/V - P_{in} \gg 50\,\mathrm{J\,cm^{-3}}$. This criterion of structuredness may also be applied to a solvent mixture, for which the internal pressure should be the mole fraction–weighted average of the P_{in} values of the components, obtained from their isobaric expansivity and the isothermal compressibility: $P_{in} = T[x_A(\alpha_{PA}/\kappa_{TA}) + x_B(\alpha_{PB}/\kappa_{TB})] - P$.

According to Marcus, a measure of the *order* of a solvent is the heat capacity density, i.e., the isobaric heat capacity, corrected for the internal heat capacity of the solvent molecule in the ideal gas state, per unit volume [2,24]. This measure, $[C_P(l) - C_P(\mathrm{i.g.})]/V$, where l denotes the liquid and (i.g.) denotes the ideal gas, has a value of > 0.6 J K^{-1} cm^{-3} for a neat *structured solvent*. This concept depends on the extra energy per unit

TABLE 2.21B Surface Tension σ at 298.15 K of Mixtures of Tetrachloromethane ($\sigma = 26.1\,\mathrm{mN\,m^{-1}}$) with Cosolvents at the Stated Mole Fractions of the Cosolvent

Cosolvent	Ref.	0.15	0.35	0.50	0.65	0.85	1.00
Hexane							
Cyclohexane							
Benzene	T	26.66	27.20	27.52	27.79	28.06	28.19
Toluene							
Chloroform	T	25.82	25.55	25.53	25.57	26.12	26.64
Methanol	e	26.94	25.44	24.48	23.64	22.73	22.21
Ethanol	e	26.30	24.14	22.81	21.71	20.63	20.09
Ethyl ether							
Tetrahydrofuran							
Dioxane							
Acetone							
Ethyl acetate	T	25.57	24.88	24.43	24.05	23.64	23.40
Pyridine							
Nitromethane							
Nitrobenzene	T	27.33	29.51	31.76	34.53	39.03	43.01
Acetonitrile	f	26.4	26.6	26.8	27.1	27.6	28.66
Dimethylformamide							
Dimethylsulfoxide							

Refs. can be found on pg. 94.

volume required to partly destroy the order when the temperature is being raised, above the energy that goes into the internal degrees of freedom of the molecules. This should be applicable to solvent mixtures too, the heat capacities and volume being the mole fraction–weighted averages of the corresponding values for the components, with $C_P(l)$ and V also including the excess heat capacity and volume.

Less readily applicable to mixtures is the concept of the entropy deficit [24], which for neat solvents also measures the amount of order in the liquid. For the neat solvent the molar entropy of vaporization is

$$\Delta_V S = \Delta_V H / T - R \ln(p/P^\circ) \qquad (2.51)$$

where p is the vapor pressure and $P^\circ = 0.1$ MPa is the standard pressure. The entropy deficit is measured against the entropy of a nonstructured solvent, an alkane with a similar carbon atom skeleton. Nonideality in the vapor phase is also taken into account by considering the derivatives of the second virial coefficients, B, with respect to the temperature.

TABLE 2.21C   Surface Tension σ at 298.15 K of Mixtures of Methanol ($\sigma = 22.3 \, \text{mN} \, \text{m}^{-1}$) with Cosolvents at the Stated Mole Fractions of the Cosolvent

Cosolvent	Ref.	0.15	0.35	0.50	0.65	0.85	1.00
Hexane				Miscibility gap			
Cyclohexane				Miscibility gap			
Benzene	T	23.50	24.87	25.79	26.63	27.6	28.23
Toluene	e	22.65	23.44	24.21	25.12	26.58	27.83
Chloroform	T	23.29	24.33	24.99	25.54	26.12	26.43
Tetrachloromethane	e	22.73	23.64	24.48	25.44	26.94	28.20
Ethanol							
Ethyl ether							
Tetrahydrofuran							
Dioxane							
Acetone	T	21.85	22.19	22.34	22.39	22.33	22.18
Ethyl acetate							
Pyridine							
Nitromethane							
Nitrobenzene							
Acetonitrile							
Dimethylformamide							
Dimethylsulfoxide	g	25.43	29.88	33.14	36.26	40.08	42.70

Refs. can be found on pg. 94.

The resulting nondimensional quantity is

$$\Delta\Delta_V S/R = [\Delta_V S_{\text{solvent}} - \Delta_V S_{\text{alkane}}]/R + (P^\circ/R)d[B_{\text{solvent}} - B_{\text{alkane}}]/dT \tag{2.52}$$

Neat solvents with $\Delta\Delta_V S/R > 2$ are structured. It appears to be difficult to apply this measure to a mixtures, because no alkane with a carbon atom skeleton that simulates the weighted average between the neat solvents can be readily defined.

A measure of structuredness that can possibly be applied to solvent mixtures where at least one of the components is polar is the *Kirkwood dipole orientation correlation parameter g*. For neat solvents this parameter is

$$g = (9k_B\varepsilon_0/4\pi N_{AV})VT\mu^{-2}(\varepsilon - 1.1n_D^2)(2\varepsilon + 1.1n_D^2)/\varepsilon(2 + 1.1n_D^2)^2 \tag{2.53}$$

TABLE 2.21D Surface Tension σ at 298.15 K of Mixtures of Ethanol ($\sigma =$ 21.9 mN m^{-1}) with Cosolvents at the Stated Mole Fractions of the Cosolvent

Cosolvent	Ref.	0.15	0.35	0.50	0.65	0.85	1.00
Hexane	h	20.61	19.14	18.14	17.36	17.07	17.81
Cyclohexane							
Benzene	T	22.99	24.12	24.96	25.80	26.93	27.77
Toluene	e	21.05	22.39	23.49	24.69	26.42	27.83
Chloroform	T	22.90	23.90	24.50	25.20	26.10	26.60
Tetrachloromethane	e	20.63	21.71	22.81	24.14	26.30	28.19
Methanol							
Ethyl ether							
Tetrahydrofuran							
Dioxane							
Acetone	T	22.95	22.81	22.66	22.47	22.16	21.89
Ethyl acetate							
Pyridine							
Nitromethane							
Nitrobenzene							
Acetonitrile (20°C)	i	23.30	23.77	24.14	24.84	26.76	29.29
Dimethylformamide							
Dimethylsulfoxide	g	23.81	26.65	29.16	32.22	37.52	42.70

Refs. can be found on pg. 94.

where k_B is Boltzmann's constant, N_{AV} is Avogadro's number, ε_0 is the permittivity of vacuum, and μ is the dipole moment of the solvent. The factor 1.1 corrects the square of the D-line refractive index n_D to yield the relative permittivity at infinite frequency (i.e., $1.1\ n_D^2 \approx \varepsilon_\infty$). The values found for structured solvents are $g > 1.7$. Whereas the molar volume V, the relative permittivity ε, and the refractive index n_D of the mixture are readily prorated for the composition, this is not straightforward for the dipole moment. This is particularly so when one of the components of the mixture is nonpolar, so that it is impossible to speak of the "average polarity" of a solvent mixture in terms of the dipole moments.

The openness of a solvent is measured by the difference between the actual molar volume and the *intrinsic volume*, which is the product of N_{AV} and the actual volume of the molecules of the solvent. This difference measures the *void volume*, and the fractional void volume, $1 - V_{intrinsic}/V$, plays an important role for, say, the viscosity of the

TABLE 2.21E   Surface Tension σ at 298.15 K of Mixtures of Acetone ($\sigma = 22.7 \, mN \, m^{-1}$) with Cosolvents at the Stated Mole Fractions of the Cosolvent

Cosolvent	Ref.	0.15	0.35	0.50	0.65	0.85	1.00
Hexane							
Cyclohexane							
Benzene	T	23.67	24.87	25.71	26.51	27.51	28.21
Toluene							
Tetrachloromethane							
Chloroform	T	23.56	24.47	24.88	25.19	25.51	25.67
Methanol	T	22.33	22.39	22.34	22.19	21.85	21.49
Ethanol	T	22.16	22.47	22.66	22.81	22.95	23.02
Ethyl ether	T	18.80	17.15	16.14	16.31	14.48	14.09
Tetrahydrofuran							
Dioxane							
Ethyl acetate							
Pyridine							
Nitromethane							
Nitrobenzene							
Acetonitrile							
Dimethylformamide							
Dimethylsulfoxide	j	24.34	26.96	29.62	32.89	38.18	42.85

Refs. can be found on pg. 94.

solvent. This measure can be applied to solvent mixtures too, but for both neat solvents and mixtures the intrinsic volume is not uniquely defined. Two of the common measures of the intrinsic volume are the McGowan [25] V_x and the van der Waals volume V_{vdW}. The former depends on just the kind and number of atoms of kind k in the molecules of the solvents, N_k, and the number of bonds between the atoms, N_{bonds}, not distinguishing between single and multiple bonds or between geometrical or positional isomers.

$$V_X / cm^3 \, mol^{-1} = \sum_k N_k V_{Xk} - 6.56 N_{bonds} \qquad (2.54)$$

The volumes assigned to the atoms, V_{Xk}, are (in $cm^3 \, mol^{-1}$): 16.35 for C, 8.71 for H, 12.43 for O, 14.39 for N, 10.48 for F, 20.95 for Cl, 26.21 for Br, 34.53 for I, 22.91 for S, and 24.87 for P. For a mixture the average intrinsic volume should be prorated according to the composition. The van der Waals intrinsic volume, V_{vdW}, also takes into account the

TABLE 2.21F   Surface Tension σ at 298.15 K of Mixtures of Acetonitrile ($\sigma = 28.3\,\mathrm{mN\,m^{-1}}$) with Cosolvents at the Stated Mole Fractions of the Cosolvent

Cosolvent	Ref.	0.15	0.35	0.50	0.65	0.85	1.00
Hexane							
Cyclohexane							
Benzene							
Toluene							
Tetrachloromethane	f	27.6	27.1	26.8	26.6	26.4	26.31
Chloroform							
Methanol							
Ethanol (20°C)	i	26.76	24.84	24.14	23.77	23.30	22.61
Ethyl ether							
Tetrahydrofuran							
Dioxane							
Acetone							
Ethyl acetate							
Pyridine							
Nitromethane							
Nitrobenzene							
Dimethylformamide							
Dimethylsulfoxide							

Refs. can be found on pg. 94.

overlap of spheres assigned to the bonded atoms and structural features (rings, branching, etc.) and can be calculated from group contributions [26]. For individual solvents the van der Waals intrinsic volumes are approximately (within $\pm 1\,\mathrm{cm^3\,mol^{-1}}$) linearly related to the McGowan intrinsic volumes [27]:

$$V_{\mathrm{vdW}}/\mathrm{cm^3\,mol^{-1}} = 1.8 + 0.674(V_{\mathrm{X}}/\mathrm{cm^3\,mol^{-1}} \qquad (2.55)$$

the uncertainty arising mainly from the independence of V_{X} from the structural features. For a mixture, again, the average van der Waals intrinsic volume should be prorated according to the composition.

The electrical conductivity of organic solvents, including that of water, and of binary mixtures of them is generally very low, but whatever conductivity does exist is due to *autoprotolysis* (if electrolyte impurities are strictly excluded). The autoprotolysis has been determined for many of the pure solvents that are miscible with water (see Ref. 2) as well as for the binary aqueous mixtures [28,29]. The relevant equilibria (with the

TABLE 2.21G   Surface Tension σ at 298.15 K of Mixtures of Nitromethane (σ = 32.1 mN m^{-1}) with Cosolvents at the Stated Mole Fractions of the Cosolvent

Cosolvent	Ref.	0.15	0.35	0.50	0.65	0.85	1.00
Hexane				Miscibility gap			
Cyclohexane				Miscibility gap			
Benzene	d	33.2	30.9	29.7	29.0	28.8	28.3
Toluene							
Tetrachloromethane							
Chloroform							
Methanol							
Ethanol							
Ethyl ether							
Tetrahydrofuran							
Dioxane							
Acetone							
Ethyl acetate							
Pyridine							
Nitromethane							
Nitrobenzene							
Dimethylformamide							
Dimethylsulfoxide							

[a]RL Schmidt, JC Randall, HL Clever. J Phys Chem 70:3912, 1966.
[b]TV Lam, GC Benson. Can J Chem 48:3773, 1970.
[c]WE Shipp. J Chem Eng Data 15:308, 1970.
[d]HP Meissner, AS Michaels. Ind Eng Chem 41:2782, 1949.
[e]SS Shastri, AK Mukherjee, TR Das. J Chem Eng Data 38:399, 1993.
[f]PI Teixeira, BS Almeida, MM Telo de Gama, JA Rueda, RG Rubio. J Phys Chem 96:8488, 1992.
[g]CM Kinart, WJ Kinart, A Bald. Phys Chem Liq 37:317, 1999.
[h]D Papaioannou, CG Panayiotou. J Chem Eng Data 39:457, 1994.
[i]DR Lide, ed. Handbook of Chemistry and Physics. 64th ed. Baton Rouge: CRC Press, 1984, p F-26.
[j]CM Kinart, WJ Kinart. Phys Chem Liq 29:1, 1995.
[T]J Timmermans. The Physico-chemical Constants of Binary Systems in Concentrated Solutions. Vols. 1 and 2. New York; Interscience, 1959.

symbols for the equilibrium constants in parentheses) according to Roses et al. [28] are $H_2O + H_2O \rightleftharpoons H_3O^+ + OH^-$ (K_W), $H_2O + HS \rightleftharpoons H_3O^+ + S^-$ (K_{bW}), $H_2O + HS \rightleftharpoons H_2S^+ + OH^-$ (K_{aW}), and $HS + HS \rightleftharpoons H_2S^+ + SH^-$ (K_{HS}). The cosolvent is written as if it were protic (HS), even if aprotic solvents are employed. The relationship between the equilibrium

TABLE 2.22 Pseudocritical Temperatures, T_{cp}/K, of Some Binary Solvent Mixtures

Cosolvent	T_{ct}/K	C_6H_6	CCl_4	MeOH	EtOH	Me_2CO	MeCN	$MeNO_2$
Hexane	508	532	529	491	502	505	524	539
Cyclohexane	554	559	556	521	529	530	550	568
Benzene	562		561	530	536	536	555	575
Toluene	592	577	574	536	546	547	568	587
Chloroform	536	550	547	521	526	523	542	563
Tetrachloromethane	556	561		528	534	533	552	572
Methanol	513	530	528		513	505	523	546
Ethanol	514	536	534	513		511	529	550
Ethyl ether	467	513	510	480	487	489	505	522
Tetrahydrofuran	540	552	549	522	527	525	544	565
Dioxane	587	576	573	542	548	547	567	588
Acetone	508	536	533	505	511		528	547
Ethyl acetate	523	543	540	506	515	516	535	552
Pyridine	617	590	587	558	563	561	581	604
Nitromethane	588	575	572	546	550	547	567	
Nitrobenzene	721	633	629	583	596	600	623	642
Acetonitrile	546	555	552	523	529	528		567
Dimethylformamide	597	581	577	546	552	551	572	593
Dimethylsulfoxide	729	640	637	605	610	608	630	655

Source: Calculated according to Ref. 22.

constants: $K_{aW} K_{bW} = K_W K_{HS}$ reduces the number of constants required for a description of the autoprotolysis equilibrium to three. In the binary mixture, the autoprotolysis constant K_{ap} is given by

$$K_{ap} = K_W x_W^2 + (K_{aW} + K_{bW})x_W x_{HS} + K_{HS} x_{HS}^2 \qquad (2.56)$$

Of the two constants, K_{aW} and K_{aW}, one generally predominates, depending on the relative acid–base strengths of water and the cosolvent, and the sum is abbreviated K_{abW}. Furthermore, the three pKs have been found experimentally to be linear with the composition: $pK_W = pK_{W(W)}$ $x_W + pK_{W(HS)}x_{HS}$, $pK_{abW} = pK_{abW(W)}x_W + pK_{abW(HS)}x_{HS}$, and $pK_{HS} = pK_{HS(W)}x_W + pK_{HS(HS)}x_{HS}$, where $pK_{W(W)}$ is the pK_{ap} of water in water (14.00 at 25°C), $pK_{W(HS)}$ is the hypothetical pK_{ap} of trace water in neat HS, etc. Therefore, five additional constants are necessary for the calculation of the autoprotolysis constant in the binary aqueous solutions, unless one or two of the terms in Eq. (2.56) can be neglected. The values of the

TABLE 2.23 pK Values for Water and the Cosolvent for the Calculation of the Autoprotolysis Constant K_{ap} for Aqueous Mixtures with Cosolvents at 298.15 K According to Eq. (2.56)

Cosolvent	p$K_{W(HS)}$	p$K_{abW(W)}$	p$K_{abW(HS)}$	p$K_{HS(W)}$[a]	p$K_{HS(HS)}$[a]
Methanol	20.82	12.98	14.56	15.09	16.91
Ethanol	22.56	13.16	16.92	15.90	19.10
n-Propanol	24.52	13.10	17.77	16.10	19.40
i-Propanol	25.47	13.27	18.53		21.08
t-Butanol	32.88	12.96	21.25	19.00	26.80
1,2-Ethanediol	19.43	12.52	13.92	15.07	15.84
Glycerol	21.86	12.24	13.44	14.10	
Ethanolamine (293 K)		2.56	6.26	[9.5]	5.14
Tetrahydrofuran	28.48	13.09	21.19	[14.8]	35.50
1,4-Dioxane	38.80	12.71	25.96	[15.6]	
Acetone	23.52	13.57	20.28	24.20	32.50
Formic acid	25.98	1.75	3.30	3.75	5.77
Acetonitrile	19.96	13.80	18.58	[24.1]	32.20
Formamide		10.43	13.11	[15.2]	16.80
N-Methylformamide	25.86	12.64	18.57	[15.6]	10.74
N,N-Dimethylformamide	27.44	12.14	22.18	[15.6]	23.10
N-Methylacetamide	24.03	12.35	19.03	[16.5]	
N,N-Dimethylacetamide	22.52	12.09	21.18		23.95
Dimethylsulfoxide	25.52	13.57	23.33	[15.5]	31.80

[a]From Ref. 2, the values of p$K_{HS(W)}$ are the p$K_{aHS(W)}$ or [p$K_{bHS(W)}$] for dilute HS in water, on the mol L^{-1} scale.
Source: Ref. 28.

required constants for many aqueous cosolvent mixtures are shown in Table 2.23 as far as available. The autoprotolysis constants, pK_{ap}, of various aqueous amide mixtures reported by Asuero et al. [29] are shown in Fig. 2.7 as functions of the composition.

Other chemical properties of solvent mixtures include their polarity, ability to donate and accept unshared electron pairs to form coordinate bonds with them, and ability to donate and accept hydrogen bonds to and from solutes and within the binary mixtures themselves. The quantities that describe these properties are generally measured by means of chemical probes, often solvatochromic probes. Such probes are molecules that change their color [expressed as the wavelength of the lowest energy peak in the ultraviolet (UV)–visible spectrum] when the polarity or the electron-pair or hydrogen-bonding donicity or acceptance changes as the

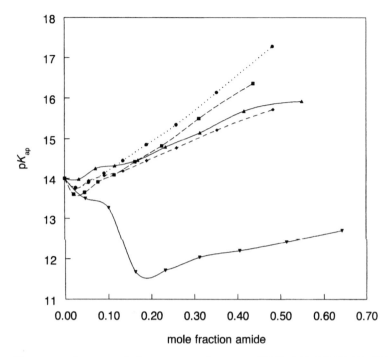

FIGURE 2.7 Autoprotolysis constants of aqueous DMF (●), NMF (▲), FA (▼), N,N-dimethylacetamide (■), and N-methylacetamide (◆), according to Asuero et al. [29].

nature and composition of the solvent or solvent mixture are varied. Other probes may utilize different spectroscopic properties, such as the NMR chemical shift, to describe the previously mentioned properties of solvents and solvent mixtures.

The question then arises whether a given probe can truly represent the behavior of the solvent mixture toward any desired solute, i.e., be the stand-in for a "general solute." Unless established otherwise, any particular probe employed would show interactions specific for it in a solvent mixture due to preferential solvation by one component of the mixture. This aspect has been the subject of only a few studies, such as those by Marcus [30] or Reta et al. [31], and no generally satisfactory answer to the question can be given. Nevertheless, a working hypothesis is that if several structurally different probes provide similar numerical values for a given property, within a reasonable margin of variation, such values can be taken to represent properly the property of the mixed solvent. This procedure introduces some vagueness into the concept of "chemical property of a solvent mixture" but is useful in correlation analysis of various modes of chemical behavior (solubility, reactivity, etc.) of solutes in series of solvents and solvent mixtures.

A widely employed measure of solvent polarity is Dimroth and Reichardt's $E_T(30)$ [32], said to represent the "general polarity" of a solvent but shown by Marcus [33] to involve mainly its hydrogen bond donation ability beside its polarity and polarizability. This is generally expressed in kcal mol^{-1} (1 kcal mol^{-1} = 4.184 kJ mol^{-1}) and is obtained from $E_T(30) = 28590/(\lambda_{max}/nm)$, where λ_{max} is the wavelength of the lowest energy peak in the spectrum of the particular 4-(2,4,6-triphenylpyridinium)-2,4-diphenylphenolate betaine employed. This polarity index is often used in the normalized dimensionless form:

$$E_T^N = [E_T(30) - 30.7]/32.4 \tag{2.57}$$

where 30.7 is the $E_T(30)$ of tetramethylsilane, the lowest measured, and 32.4 is the difference between the $E_T(30)$ of water, 63.1, the highest measured, and that of tetramethylsilane [32]. The $E_T(30)$ and E_T^N have been shown *not* to be linear functions of the composition in a mixed solvent, and various expressions have been proposed to describe the deviations from linearity.

An empirical expression applicable to binary mixtures due to Langhals [34] is

$$E_T(30) = E_T(30)_B + E_D \ln[(c_A/c^*) + 1] \tag{2.58}$$

where solvent B is the less polar one [the one with the lower $E_T(30)$ value], c_A is the concentration (in moles per liter) of the more polar component, A, and E_D and c^* are parameters specific for each pair of solvents. This expression was applied to a large number of mixtures and in some cases could not cover the entire composition range and had to be split into two regions (with different E_D and c^* values). It does not convey any information concerning possible preferential solvation (of the probe). Table 2.24 shows the E_D and c^* parameters for mixtures involving some selected solvents. More informative about possible preferential solvation (see Section 4.4) is the approach, used among others by Dawber et al. [34a], that defines the excess polarity $\Delta E_T(30)$, i.e.,

$$\Delta E_T(30) = E_T(30) - [x_A E_T(30)_A + x_B E_T(30)_B] \tag{2.59}$$

TABLE 2.24   Pairs of c^*, E_D Parameters of Eq. (2.58) for Binary Solvent Mixtures A + B, Where A Is the More Polar Solvent

Solvent B	$E_T(30)_B$	Water	Methanol	Ethanol	Acetone	Aceto-nitrile	Nitro-methane
$E_T(30)_A$		63.1	55.4	51.9	42.2	46.0	46.3
Methanol	55.4	24.0, 3.91					
Ethanol	51.9	6.83, 1.64					
t-Butanol	43.9	1.01, 2.82					
Dioxane	36.0	0.58, 4.34	0.98, 7.67	0.72, 4.99		0.77, 3.23	1.01, 3.49
Acetone	42.2	0.31, 2.83	0.10, 2.53	0.14, 2.27			
Pyridine	40.2	5.48, 7.09	5.84, 6.92	12.8, 9.64	32.1, 4.01		13.6, 6.46
MeCN	46.0	0.15, 2.07	0.06, 1.83	0.10, 1.83			
DMF	43.8	11.4, 9.24	0.70, 3.55	0.61, 2.78			
DMSO	45.0				2.69, 3.95		
Nitro-methane	46.3		0.01, 1.66	0.03, 1.41			

Source: Ref. 34.

of the mixture. Positive values of $\Delta E_T(30)$ denote preferential solvation of the probe by the more polar solvent; negative values denote that the less polar one preferentially solvates it. The use of $\Delta E_T(30)$ values of mixtures has been sufficiently widespread to merit the recording of the $\Delta E_T(30)$ values reported to date. These are to be found in Table 2.25. It should be noted that for practically all mixtures where water is one of the components $\Delta E_T(30) < 0$ with few exceptions (cosolvent-rich mixtures of water with tetrahydrofuran, 1,4-dioxane, acetone, and acetonitrile). For these exceptions the curves are strongly S-shaped, $\Delta E_T(30)$ being negative for water-rich and positive for cosolvent-rich mixtures. Such behavior has been interpreted in terms of the microheterogeneity of the mixtures. For nonaqueous mixtures generally $\Delta E_T(30) > 0$. All this may tell us something about the preferential solvation of the particular probe (the penta-phenyl pyridinium phenolate betaine) used to determine $E_T(30)$ but not necessarily anything about how other solutes behave in the solvent mixture under investigation. This point is discussed further in Section 4.4. The properties of this betaine and a similar one with two chlorine atoms instead of phenyl groups on the phenolate ring in mixed solvents with regard to preferential solvation were further studied by Tada et al. [34b].

The Kamlet–Taft [35] polarity–polarizability parameter π^* has been determined for much fewer binary solvent mixtures than $E_T(30)$

TABLE 2.25A   Deviation $\Delta E_T(30)$ for Mixtures of Solvent A, Water $[E_T(30) = 63.1]$ with Solvents B at Various Mole Fractions x_B at the Heads of Columns

Solvent B	$E_T(30)_B$	0.15	0.35	0.50	0.65	0.85
Methanol	55.7	−1.8	−2.4	−2.1	−1.6	−0.9
Ethanol	51.7	−3.1	−4.1	−3.5	−2.3	−0.8
n-Propanol	50.6	−5.4	−5.3	−3.3	−2.0	−1.2
i-Propanol	48.7	−6.3	−6.2	−4.3	−2.9	−2.2
t-Butanol	43.9	−3.2	−5.6	−2.7	−1.1	−0.2
1,2-Ethanediol	56.3	−1.4	−1.8	−1.7	−1.4	−0.7
1,2-Dimethoxyethane[a]	38.2	−6.2	−5.0	−2.9	−1.3	+0.4
Tetrahydrofuran	37.5	−6.7	−4.3	−0.6	+1.7	+1.3
1,4-Dioxane	36.0	−5.0	−3.7	−0.9	+1.4	+2.1
Pyridine	40.3	−5.7	−4.5	−4.2	−1.5	−0.7
Piperidine	35.5	−5.7	−5.6	−3.7	−1.8	−0.4
Acetone	42.2	−3.9	−2.8	−0.3	+1.5	+1.7
Acetonitrile	46.0	−1.9	−1.0	+1.0	+3.2	+3.8
Dimethylformamide	43.8	−3.8	−4.5	−3.6	−2.4	−1.1
N-Methylpyrrolidin-2-one[b]	42.2	−4.9	−5.3	−4.8	−3.9	−1.7
Dimethylsulfoxide	45.0	−3.5	−4.6	−4.1	−3.2	−1.7

Source: Calculated from data in Ref. 30 unless otherwise noted.
Footnotes can be found on pg 106.

values. This quantity takes into account both the polarity of the solvent, which can be expressed through its dipole moment μ_i or molar polarization, P_i, Eq. (2.7), and its polarizability, which can be expressed through its refractive index, n_{Di}, or its molar refraction, R_{Di}, Eq. (2.9). For neat solvents it was shown to be a linear combination of these quantities [2,36]. Because the values of π^* have been normalized to cover the range 0 to 1 (but $\pi^* = 1.09$ for water), the deviations from linearity with the composition

$$\Delta\pi^* = \pi^* - (x_A\pi^*_A + x_B\pi^*_B) \qquad (2.60)$$

are numerically small. Furthermore, since π^* is generally an average quantity, based on the use of several solvatochromic probes, its probable error for neat solvents is ± 0.02. Hence, values of $|\Delta\pi^*| < 0.05$ have little significance. The existing data are summarized in Table 2.26. For many mixtures nearly linear dependence (within 0.05 units) of π^* on the mole fraction composition is noted, but there are exceptions, in particular when one component has a high π^* value and the other a low one.

TABLE 2.25B   Deviation $\Delta E_T(30)$ for Mixtures of Solvent A, Methanol $[E_T(30) = 55.4]$ with Solvents B at Various Mole Fractions x_B at the Heads of Columns

Solvent B	$E_T(30)_B$	0.15	0.35	0.50	0.65	0.85
Benzene	34.5	+1.0	+3.8	+6.4	+8.1	+6.6
Toluene	34.7	+0.4	+2.3	+4.7	+7.7	+6.5
Chloroform	39.1	+1.0	+2.4	+3.4	+4.1	+3.7
Ethanol	51.7	−0.3	−0.3	−0.3	−0.3	−0.3
2-Butanol[c]	47.2	+0.3	+0.1	+0.3	−0.2	0.1
t-Pentanol[c]	41.1	+2.6	+3.1	+2.1	+1.1	−0.3
1,2-Ethanediol	56.9	+0.2	+0.2	+0.2	+0.2	+0.2
Tetrahydrofuran	37.5	+1.6	+3.3	+5.1	+6.7	+5.8
1,4-Dioxane	36.0	+1.6	+3.2	+4.6	+6.0	+5.3
Acetone	42.2	+2.2	+4.1	+5.4	+6.8	+5.7
Pyridine	40.2	+0.1	−0.6	+0.7	+1.7	+1.4
Acetonitrile	46.0	+1.6	+3.1	+4.5	+6.0	+6.2
Nitromethane	46.3	+5.7	+5.8	+6.5	+7.5	+7.8
Dimethylformamide	43.8	+2.6	+3.5	+3.6	+3.8	+3.1
Dimethylsulfoxide	45.0	+0.6	+1.1	+1.6	+1.8	+1.5

Source: Calculated from data in Ref. 30 unless otherwise noted.
Footnotes can be found on pg. 106.

The electron pair donation (Lewis basicity, or electron density donation for solvents such as benzene that are not normally speaking Lewis bases) or the hydrogen bond acceptance abilities of solvent mixtures have been reported somewhat more extensively than the polarity–polarizability. The expression for the excess Kamlet–Taft β parameter [35]:

$$\Delta\beta = \beta - (x_A\beta_A + x_B\beta_B) \tag{2.61}$$

is analogous to that for the excess polarity–polarizability. Again, because β is generally an average quantity, based on the use of several solvatochromic probes, its probable error for neat solvents is ± 0.02. Values for mixtures were generally obtained from a single probe only. This was often the Koppel B_K [37], obtained from infrared wavenumber shifts of O–D in deuterated methanol or of O–H in phenol, related to β by the expression $\beta = 0.975 B_K + 0.002$. Hence, values of $|\Delta\beta| < 0.05$ have little significance. The values of $\Delta\beta$ for aqueous mixtures are shown in Table 2.27a and for mixture of methanol and other nonaqueous solvents with cosolvents in Table 2.27b and 2.27c. It should be noted that for mixtures involving a solvent, A, with high donicity β and another, B,

TABLE 2.25C Deviation $\Delta E_T(30)$ for Mixtures of Solvent A, Ethanol [$E_T(30) = 51.9$] with Solvents B at Various Mole Fractions x_B at the Heads of Columns

Solvent B	$E_T(30)_B$	0.15	0.35	0.50	0.65	0.85
c-Hexane	30.9	+1.7	+4.4	+7.9	+11.5	+11.7
Benzene	34.5	+1.2	+3.3	+5.3	+7.0	+6.6
Chloroform	39.1	+0.6	+1.4	+2.3	+3.3	+3.2
t-Butanol	43.7	+0.4	+0.8	+1.1	+1.5	+1.5
2,2,2-Trifluoroethanol[c]	59.5	+1.2	+1.6	+1.6	+1.3	+0.6
1,2-Ethanediol	56.9	+0.8	+0.5	0	−0.3	−0.5
1,4-Dioxane	36.0	+1.4	+2.9	+4.0	+4.7	+3.9
Acetone	42.2	+1.4	+2.9	+3.9	+4.7	+4.1
Ethyl acetate[d]	38.1	+1.6	+3.2	+4.8	+5.9	+7.4
Pyridine	40.2	−2.9	−2.4	−1.3	−0.7	−0.5
Acetonitrile	46.0	+2.7	+3.8	+4.8	+5.5	+5.9
Nitromethane	46.3	+5.4	+4.9	+5.0	+5.5	+6.0
Dimethylformamide	43.8	+1.9	+2.6	+3.0	+3.4	+2.8
Dimethylsulfoxide	45.0	+1.3	+1.7	+1.7	+1.4	+0.8

Source: Calculated from data in Ref. 30 unless otherwise noted.
Footnotes can be found on pg. 106.

with low donicity, the donicity of the mixture remains near the value of A until quite large concentrations of B in the mixture are reached. Hence, $\Delta\beta$ for such mixtures tends to increase toward large positive values as x_B increases.

The ability of a solvent to donate a hydrogen atom toward the formation of a hydrogen bond (Lewis acidity) is measured by its Kamlet–Taft α value [35]. As mentioned earlier, the $E_T(30)$ value of a protic solvent is linearly dependent on α with an additional contribution from the polarity–polarizability, π^*. Hence the deviations of α from a linear dependence on the solvent composition can be given by

$$\Delta\alpha = 0.0657\Delta E_T(30) - 0.757\Delta\pi^* \qquad (2.62)$$

There are other probes that measure α which is then obtained as an average value of the results of all the probes. However, the mutual consistency of various probes is generally not better than ± 0.08 units, on a scale that goes from 0 for aprotic and nonprotogenic solvents to 1.17 for water and nearly up to 2 for a solvent such as 1,1,1,3,3,3-hexafluoro-2-

TABLE 2.25D Deviation $\Delta E_T(30)$ for Mixtures of Solvent A, Acetone $[E_T(30) = 42.2]$ with Solvents B at Various Mole Fractions x_B at the Heads of Columns

Solvent B	$E_T(30)_B$	0.15	0.35	0.50	0.65	0.85
Toluene[a]	33.9	+0.6	+1.1	+1.5	+2.0	+2.8
Tetrachloromethane	32.5	+0.7	+1.8	+2.4	+3.0	+2.7
Chloroform	39.1	+1.7	+2.0	+2.0	+1.5	+1.0
Dichloromethane[e]	41.1	+0.8	+1.0	+1.1	+1.1	+0.8
Methanol	55.4	+5.7	+6.8	+5.4	+4.1	+2.2
Ethanol	51.9	+4.1	+4.7	+3.9	+2.9	+1.4
i-Propanol	49.7	+0.5	+1.3	+1.2	+0.9	+0.4
t-Butanol	43.7	−1.4	−0.9	−0.6	−0.5	+0.8
Trimethyl phosphate[f]	43.5	+0.2	+0.3	+0.4	+0.3	+0.2
Pyridine	40.2	−0.6	−0.4	−0.2	−0.1	−0.1
Acetonitrile	45.6	+0.4	+0.7	+0.8	+0.5	+0.3
Nitromethane	46.3	+0.5	+0.7	+0.6	+0.6	+0.4
N-methylpyrrolidin-2-one[b]	42.2	+0.5	+0.6	+0.6	+0.6	+0.3
Dimethylsulfoxide	45.0	+0.5	+0.8	+0.8	+0.7	+0.4

Source: Calculated from data in Ref. 30 unless otherwise noted.
Footnotes can be found on pg. 106.

propanol. One of the probes that has been used for measuring the Lewis acidity is 1-ethyl-4-cyanopyridine, yielding Kosower Z values [38] in kcal mol^{-1} (1 kcal mol^{-1} = 4.184 kJ mol^{-1}). These, for neat solvents and presumably also for solvent mixtures, are related to α by $\alpha = 0.0485Z - 2.75 - 0.46\pi^*$. Hence:

$$\Delta\alpha = 0.0485[Z - (x_A Z_A + x_B Z_B)] - 0.46\Delta\pi^*$$
$$\approx 0.0485[Z - (x_A Z_A + x_B Z_B)] \qquad (2.63)$$

the approximate equality being required because the π^* values for the mixed binary solvents for which Z values are available are not known and $\Delta\pi^*$ tends to have small values anyway. The resulting $\Delta\alpha$ values are shown in Table 2.28a for mixtures involving water as solvent A with various cosolvents B and in Table 2.28b for completely nonaqueous mixtures. Most of the available α values of mixtures are due to work with just one probe, so the uncertainty due to the averaging of values from two or more probes, that of ±0.08 noted for neat solvents, does not apply. On the other hand, the uncertainties in the correlations of α with $E_T(30)$ and with Z are themselves fairly large, so that $|\Delta\alpha| < 0.05$ is probably

TABLE 2.25E Deviation $\Delta E_T(30)$ for Mixtures of Solvent A, Acetonitrile $[E_T(30) = 45.6]$ with Solvents B at Various Mole Fractions x_B at the Heads of Columns

Solvent B	$E_T(30)_B$	0.15	0.35	0.50	0.65	0.85
Benzene	34.5	+0.9	+2.0	+2.7	+3.0	+2.1
Methanol	55.4	+6.2	+6.0	+4.5	+3.1	+1.6
Ethanol	51.9	+5.9	+5.5	+4.8	+3.8	+2.7
i-Propanol	49.7	+4.0	+3.8	+2.5	+1.7	+0.8
t-Butanol	43.7	+2.8	+3.1	+2.6	+2.1	+1.4
1,4-dioxane	36.0	+1.5	+2.3	+2.8	+3.1	+2.4
Acetone[g]	42.2	+0.3	+0.5	+0.8	+0.7	+0.4
2-Butanone[g]	41.3	+0.6	+1.0	+1.0	+1.0	+0.7
Ethyl acetate[g]	38.1	+2.0	+2.2	+2.2	+1.7	+0.9
Dimethylformamide[g]	43.8	+0.1	+0.2	+0.3	+0.3	+0.2
Dimethylsulfoxide	45.0	−0.1	−0.1	−0.1	−0.1	−0.1

Source: Calculated from data in Ref. 30 unless otherwise noted.
Footnotes can be found on pg. 106.

insignificant. For solvent mixtures involving water the values of $\Delta\alpha$ are generally negative, except for the cases of dioxane, acetone, and acetonitrile, where the $\alpha(x)$ curves are strongly S-shaped, indicating microheterogeneity. For nonaqueous systems the $\Delta\alpha$ values tend to be positive, but there are exceptions.

2.5 ASSOCIATED MIXTURES

In many cases, the properties of binary solvent mixtures have been interpreted in terms of the formation of definite associated species, either self-associated for one or the other component or mutually associated between the molecules of the components. In such cases, equilibrium constants, standard Gibbs energies of association, enthalpies of association, etc. have been calculated and are capable of representing some or all of the observed thermodynamic functions. This representation is often reversed; that is, the equilibrium constants etc. are calculated from fitting the thermodynamic data by employing suitable models. Spectroscopic methods are generally used either independently or for the confirmation of this interpretation of the behavior of solvent mixtures in terms of associated species. In optimal cases spectroscopy is able to point out features

TABLE 2.25F Deviation $\Delta E_T(30)$ for Mixtures of Miscellaneous Solvents A with Solvents B at Various Mole Fractions x_B.*

Solvent A	$E_T(30)_A$	Solvent B	$E_T(30)_B$	0.15	0.35	0.50	0.65	0.85
						x_B		
Toluene[e]	33.9	THF[e]	37.5	+0.3	+0.7	+0.5	+0.3	+0.3
Toluene[e]	33.9	1,4-Dioxane[e]	36.0	0	0	+0.1	0	0
Toluene[e]	33.9	Ethyl acetate[e]	38.1	+1.2	+1.1	+0.8	+0.6	+0.3
Toluene[e]	33.9	Chloroform[e]	39.1	+2.0	+1.9	+1.7	+1.5	+0.7
Toluene[e]	33.9	DMF[e]	43.8	+3.5	+4.0	+3.2	+2.4	+1.1
Toluene[e]	33.9	Nitromethane[e]	46.3	+4.0	+3.1	+2.1	+1.1	+0.4
Tetrachloromethane[f]	32.5	Dimethylsulfoxide[f]	45.0	+7.0	+5.7	+4.4	+3.1	+1.3
Tetrachloromethane[f]	32.5	Trimethyl phosphate[f]	43.5	+4.4	+3.8	+3.1	+1.9	+0.8
Tetrachloromethane[f]	32.5	Tributyl phosphate[f]	39.6	+2.9	+2.9	+2.5	+1.8	+0.8
Chloroform[f]	39.1	Tetrachloromethane[f]	32.5	+1.1	+1.4	+1.8	+2.1	+1.6
Chloroform[f]	39.1	Trimethyl phosphate[f]	43.5	+1.2	+2.2	+2.5	+2.5	+1.6
1,2-Dibromoethane[i]	38.3	1,2-dibromopropane[i]	39.1	0.0	0.0	0.0	0.0	0.0
2-Propanol[j]	48.3	Formamide[j]	55.8	+1.4	+1.5	+1.2	+0.8	+0.3
2-Propanol[j]	48.3	N-Methylformamide[j]	54.1	+2.3	+2.5	+2.1	+1.6	+0.7
2-Propanol[j]	48.3	Dimethylformamide[j]	43.5	+1.6	+2.2	+2.4	+2.5	+1.7
2-Methyl-2-propanol[j]	43.9	Formamide[j]	55.8	+3.0	+2.9	+2.4	+1.8	+0.8
2-Methyl-2-propanol[j]	43.9	N-methylformamide[j]	54.1	+4.3	+4.3	+3.6	+2.7	+1.2
2-methyl-2-propanol[j]	43.9	Dimethylformamide[j]	43.5	+1.1	+1.8	+2.0	+1.9	+1.3
Pyridine[f]	40.2	Benzene[f]	34.5	+0.9	+1.0	+1.0	+1.1	+0.9
Nitromethane[k]	46.3	Ethyl acetate[k]	38.1	+2.1	+1.9	+2.0	+1.5	+1.1
Nitromethane[k]	46.3	Acetone[k]	42.2	+0.4	+0.6	+0.6	+0.7	+0.5
Nitromethane[k]	46.3	DMF[k]	43.8	-0.3	-0.2	-0.2	-0.1	0
Nitromethane[k]	46.3	DMSO[k]	45.1	-0.2	-0.3	-0.3	-0.3	-0.2
Nitromethane[f]	46.3	Pyridine[f]	40.2	-0.6	-0.4	-0.2	-0.1	-0.1
Nitrobenzene[h]	41.2	Benzene[h]	34.5	+2.3	+2.5	+2.3	+1.6	+1.5
N-methylformamide[g]	54.1	Formamide[g]	55.8	0	0	0	0	0

TABLE 2.25F Continued

Solvent A	$E_T(30)_A$	Solvent B	$E_T(30)_B$	x_B 0.15	0.35	0.50	0.65	0.85
Dimethylformamide[g]	43.5	Formamide[g]	55.8	+2.9	+2.9	+2.3	+1.6	+0.7
Dimethylformamide[g]	43.5	N-Methylformamide[g]	54.1	0	0	0	0	0
N-methylpyrrolidin-2-one(NMPy)[b]	42.1	Hexamethyl phosphoric triamide[b]	40.9	0	0	0	0	0
NMPy[b]	42.1	2-Propanol[b]	48.4	+1.7	+2.0	+1.9	+2.0	+1.6
NMPy[b]	42.1	1,4-Dioxane[b]	36.0	+0.7	+1.2	+1.6	+2.2	+2.0
NMPy[b]	42.1	Acetone[b]	42.2	+0.5	+0.6	+0.6	+0.6	+0.4
NMPy[b]	42.1	1,2-Dichloroethane[b]	41.3	+0.6	+1.1	+1.3	+1.3	+0.8
NMPy[b]	42.1	Dimethylsulfoxide[b]	45.1	+0.1	+0.2	+0.2	+0.1	0
Dimethylsulfoxide[f]	45.0	Benzene[f]	34.5	+0.9	+2.1	+3.4	+4.5	+4.1
Dimethylsulfoxide[f]	45.0	Chloroform[f]	39.1	+1.2	+2.1	+2.5	+2.7	+2.1
Dimethylsulfoxide[f]	45.0	CCl_4[f]	32.5	+1.7	+3.3	+4.7	+5.8	+7.1
Dimethylsulfoxide[l]	45.0	i-Propanol[l]	49.7	+0.3	+0.9	+1.1	+1.3	+0.6
Dimethylsulfoxide[l]	45.0	t-Butanol[l]	43.7	+1.1	+1.9	+2.0	+1.9	+1.1

* See Ref. 30 for further systems that have been studied.

[a]L Novaki, OA El Scoud. Ber Bunsenges Phys Chem 101:902, 1997.

[b]M Kemell, P Pirilä. J Solution Chem 29:87, 2000 (average of data at 20 and 30°C).

[c]H Elias, G Gumbel, S Neitzel, H Volz. Fresenius Z Anal Chem 306:240, 1981.

[d]D Banerjee, A Kumar Laha, S Bagchi. J Chem Soc Faraday Trans 91:631, 1995.

[e]PME Mancini, A Terenzani, MG Gaspardi, LR Vottero. J Phys Org Chem 8:617, 1995.

[f]ZB Maksimović, C Reichardt, A Spirić. Fresenius Z Anal Chem 270:100, 1974.

[g]PME Mancini, A Terenzani, C Adam, A Perez, LR Vottero. J Phys Org Chem 12:207, 1999.

[h]E Dutkiewicz, A Jakubowska, M Dutkiewicz. Spectrochim Acta A48:1409, 1992.

[i]S Balakrishnan, AJ Easteal. Aust J Chem 34:933, 1981.

[j]K Herodes, I Leito, I Koppel, M Roses. J Phys Org Chem 12:109, 1999.

[k]PME Mancini, A Terenzani, C Adam, A Perez, LR Vottero. J Phys Org Chem 12:713, 1999.

[l]IA Kopel, JB Kopel. Org React (Tartu) 20:523, 1983.

TABLE 2.26A   Deviation $\Delta\pi^*$ for Mixtures of Solvent A, Water ($\pi^* = 1.09$) with Cosolvents B at Various Mole Fractions x_B at the Heads of Columns

Solvent B	π_B^*	0.15	0.35	0.50	0.65	0.85
Methanol	0.60	0.077	0.075	0.069	0.054	0.038
Ethanol	0.54	0.138	0.035	−0.022	−0.031	0.015
i-Propanol	0.50	−0.030	−0.101	−0.080	−0.027	0.030
t-Butanol[a]	0.41	−0.176	−0.153	−0.065	−0.019	
1,2-Ethanediol[a]	0.92	0.010	0.065	0.030	0.010	
Tetrahydrofuran	0.60	−0.062	−0.138	−0.114	−0.056	0.014
1,4-Dioxane	0.54	0.000	−0.053	−0.039	−0.024	−0.004
Acetone	0.69	0.002	−0.041	−0.060	−0.056	−0.020
Formic acid	0.99	0.051	0.062	0.058	0.045	0.025
Acetic acid	0.58	0.108	0.094	0.072	0.072	0.072
Pyridine	0.89	0.085	0.110	0.072	0.043	0.027
Acetonitrile	0.75	−0.007	−0.077	−0.061	−0.044	−0.025
Formamide	1.15	0.043	0.049	0.040	0.030	0.020
Dimethylformamide	0.89	0.063	0.043	0.007	−0.011	−0.004
Dimethylsulfoxide	0.98	0.049	0.035	0.014	0.001	−0.002

Source: Calculated from data in Ref. 30 unless otherwise noted.
Footnote can be found on pg. 108.

that are unique to the assumed associated species and cannot be due to monomeric molecules of the components alone. This, then, constitutes a validation of the model employed.

Ideal mixing of the species, whether monomers or associates, is often assumed, so that the entire values of the measured excess thermodynamic quantities (deviations from Raoult's law, heats of mixing, etc.) are ascribed to the association. However, nonspecific interactions, describable by dispersion or dipole interactions or by large discrepancies between the sizes of the molecules of the two components, must generally also be taken into account. These lead to activity coefficients of the actual species present that differ from unity. On the other hand, doubts concerning the formation of the associates, if deduced solely from fitting of thermodynamic excess functions, can be raised when the association is not particularly strong or spatially well oriented (by hydrogen bonds or coordinative bonding in donor–acceptor adducts). An alternative interpretation of the observed behavior in terms of nonspecific interactions should then, and often can, be provided.

TABLE 2.26B Deviation $\Delta\pi^*$ for Mixtures of Miscellaneous Solvents A with Solvents B at Various Mole Fractions x_B

Solvent A	π^*_A	Solvent B	π^*_B	x_B 0.15	0.35	0.50	0.65	0.85
Methanol	0.75	Tetrachloromethane	0.26	0.054	0.098	0.121	0.127	0.084
t-Butanol	0.61	Tetrachloromethane	0.26	0.013	0.002	0.005	0.012	0.024
Tetrahydrofuran	0.58	Tetrachloromethane	0.26	0.029	0.061	0.073	0.068	0.038
Ethyl acetate[b]	0.55	Dichloromethane	0.82	-0.016	-0.045	-0.059	-0.069	-0.063
Ethyl acetate[b]	0.55	Chloroform	0.58	-0.022	-0.047	-0.058	-0.084	-0.102
Acetonitrile[b]	0.66	Acetone	0.62	0.084	0.090	0.093	0.071	0.071
Acetonitrile[b]	0.66	2-Butanone	0.60	0.082	0.085	0.094	0.078	0.069
Acetonitrile[b]	0.66	Ethyl acetate	0.45	0.107	0.094	0.101	0.106	0.119
Acetonitrile[b]	0.66	Dichloromethane	0.82	0.004	0.012	0.020	0.015	-0.024
Acetonitrile[b]	0.66	Chloroform	0.58	0.027	0.034	0.036	0.035	0.005
DMF	0.88	Acetonitrile[b]	0.66	0.022	0.036	0.058	0.076	0.110
Dimethylsulfoxide[b]	1.00	Dichloromethane	0.80	-0.010	-0.010	-0.008	0.000	0.011
Dimethylsulfoxide[b]	1.00	Chloroform	0.58	-0.078	-0.087	-0.110	-0.126	-0.112
Dimethylsulfoxide	1.00	Tetrachloromethane	0.26	0.107	0.101	0.148	0.253	0.307
Dimethylsulfoxide	1.00	1,2-Dichloroethane	0.79	0.001	0.004	0.005	0.006	0.009
Dimethylsulfoxide	1.00	Chlorobenzene	0.67	0.016	0.029	0.045	0.064	0.059
Dimethylsulfoxide	1.00	Acetonitrile[b]	0.66	0.037	0.071	0.074	0.079	0.115

Source: Calculated from data in Ref. 30 unless otherwise noted.
[a]RJ Sindreu, ML Moya, FS Burgos, AG Gonzalez. J Solution Chem 25:289, 1996.
[b]PME Mancini, A Terenzani, C Adam, A del Perez, LR Vottero. J Phys Org Chem 12:207, 713, 1999.

TABLE 2.27A Deviation $\Delta\beta$ for Mixtures of Solvent A, Water ($\beta = 0.47$), with Cosolvents B at Various Mole Fractions x_B at the Heads of Columns

Solvent B	β_B	0.15	0.35	0.50	0.65	0.85
Methanol	0.66	0.111	0.136	0.105	0.057	0.008
Ethanol	0.75	0.035	0.121	0.056	0.024	0.031
i-Propanol	0.84	0.108	0.103	0.074	0.059	0.066
t-Butanol[a]	0.93	0.149	0.147	0.098	0.033	
1,2-Ethanediol[a]	0.52	0.000	0.030	0.062	0.097	0.040
Tetrahydrofuran	0.55	0.193	0.148	0.137	0.111	0.076
1,4-Dioxane	0.37	0.049	0.127	0.177	0.198	0.142
Acetone	0.48	0.138	0.182	0.158	0.111	0.045
Acetic acid	0.45	0.120	0.117	0.087	0.067	0.059
Pyridine	0.64	0.045	0.083	0.097	0.094	0.059
Acetonitrile	0.40	0.150	0.129	0.124	0.122	0.083
Formamide	0.48	0.027	0.035	0.029	0.018	0.005
Dimethylformamide	0.69	0.070	0.101	0.098	0.078	0.039
Dimethylsulfoxide	0.76	0.071	0.071	0.058	0.045	0.006

Source: Calculated from data in Ref. 30 unless otherwise noted.
Footnotes can be found on pg. 112.

Suppose that the following association reactions take place in the general case of an *associated binary mixture* of the components A and B:

$$pA \rightleftharpoons A_p \qquad \text{and} \qquad qB \rightleftharpoons B_q \qquad (2.64)$$

constituting self-association and

$$kA + lB \rightleftharpoons A_k B_l \qquad (2.65)$$

representing mutual association. These equilibria have definite values of the stoichiometric coefficients p, q, k, and l that might be unity or some higher integer or sets of integers. Then the total number of moles of A in the mixture is

$$n_A = \sum_p pn_{Ap} + \sum_k kn_{A_k B_l} \qquad (2.66a)$$

and the total number of moles of B in the mixture is

$$n_B = \sum_q qn_{Bq} + \sum_l ln_{A_k B_l} \qquad (2.66b)$$

The mole fractions of the various species, A_1, A_p, B_1, B_q, $A_k B_l$, etc. are then $x_{A_1} = n_{A_1}/(n_A + n_B)$ for the monomers of A, $x_{A_p} = n_{A_p}/(n_A + n_B)$ for

TABLE 2.27B Deviation $\Delta\beta$ for Mixtures of Solvent A, Methanol ($\beta = 0.66$), with Cosolvents B at Various Mole Fractions x_B at the Heads of Columns

Solvent B	β_B	0.15	0.35	0.50	0.65	0.85
Benzene	0.10	0.018	0.120	0.211	0.268	0.220
Tetrachloromethane[c]	0.10	0.097	0.269	0.379	0.424	0.300
Chloroform	0.10	0.043	0.116	0.156	0.144	0.023
Diethyl ether	0.46	0.050	0.076	0.077	0.065	0.035
Tetrahydrofuran	0.55	0.018	0.041	0.054	0.058	0.042
1,4-Dioxane	0.37	0.025	0.071	0.105	0.123	0.096
Acetone	0.48	0.010	0.012	0.001	-0.006	-0.004
Ethyl acetate	0.45	0.031	0.061	0.068	0.063	0.035
Propylene carbonate	0.40	-0.044	-0.061	-0.052	-0.035	-0.012
Acetonitrile	0.40	-0.019	-0.027	-0.025	-0.019	-0.008
Nitromethane	0.06	-0.067	-0.089	-0.074	-0.046	-0.010
Formamide	0.48	-0.008	-0.020	-0.028	-0.032	-0.024
Dimethylformamide	0.69	0.012	0.017	0.014	0.010	0.003
Dimethylsulfoxide	0.76	-0.012	0.006	0.016	0.019	0.008

Source: Calculated from Krygowski et al.[b] data unless otherwise noted.
Footnotes can be found on pg. 112.

the oligomers of A, $x_{B_1} = n_{B_1}/(n_A + n_B)$ for the monometers of B, $x_{B_q} = n_{B_q}/(n_A + n_B)$ for the oligomers of B, $x_{A_k B_l} = n_{A_k B_l}/(n_A + n_B)$ for adducts between A and B, etc. The condition of equilibrium then requires that the chemical potentials of the nominal components equal those of the monomeric species as stressed by Prigogine and Defay [39]:

$$\mu_A = \mu_{A_1} \quad \text{and} \quad \mu_B = \mu_{B_1} \tag{2.67}$$

This result is independent of the ideality or otherwise of the mixture and of the nature of the associated species formed. The activity coefficient of component A in the solvent mixture is measurable from the vapor pressure and composition and is given by

$$f_A = x_{A_1} f_{A_1} / x_A x_{A_1}{}^0 f_{A_1}{}^0 \tag{2.68}$$

and similarly for component B. Superscript zeros on the factors in the denominator pertain to quantities characterizing the pure component, where, of course, only self-association may take place. Expression (2.68) relates the activity coefficient of the component to the mole fractions and activity coefficients of the monomers (the non-associated species).

TABLE 2.27c Deviation $\Delta\beta$ for Mixtures of Solvent A with Cosolvents B at Various Mole Fractions x_B at the Heads of Columns

Solvent A	β_A	Solvent B	β_B	x_B				
				0.15	0.35	0.50	0.65	0.85
Ethanol	0.75	Cyclohexane[d]	0.00	0.140	0.354	0.475	0.563	0.730
Ethanol	0.75	Benzene[d]	0.10	0.056	0.121	0.174	0.210	0.163
Ethanol	0.75	Acetone	0.48	0.000	-0.006	-0.011	-0.015	-0.012
Ethanol	0.75	Acetonitrile	0.40	-0.027	-0.035	-0.029	-0.017	-0.002
Ethanol	0.75	DMF	0.69	0.035	0.051	0.048	0.036	0.014
i-Propanol	0.84	Acetone	0.48	-0.019	-0.024	-0.018	-0.010	-0.001
i-Propanol	0.84	Acetonitrile	0.40	-0.047	-0.065	-0.055	-0.036	-0.009
t-Butanol	0.93	Tetrachloromethane	0.10	0.063	0.139	0.178	0.186	0.125
Tetrahydrofuran	0.55	Tetrachloromethane	0.10	0.097	0.181	0.242	0.280	0.230
Acetone[e]	0.48	Acetonitrile	0.40	0.081	0.066	0.050	0.049	0.065
2-Butanone[e]	0.48	Acetonitrile	0.40	0.124	0.110	0.094	0.067	0.069
Ethyl Acetate[e]	0.45	Dichloromethane	0.10	0.006	0.071	0.129	0.148	0.064
Ethyl acetate[e]	0.45	Chloroform	0.10	-0.091	-0.026	0.017	0.072	-0.021
Ethyl acetate[b]	0.45	Acetonitrile	0.40	0.069	0.075	0.079	0.066	0.063
Pyridine	0.64	Acetonitrile	0.40	0.159	0.170	0.193	0.206	0.147
Pyridine	0.64	Nitromethane	0.06	0.231	0.326	0.391	0.426	0.371
Acetonitrile[e]	0.40	Dichloromethane	0.10	0.058	0.094	0.106	0.113	0.066
Acetonitrile[e]	0.40	Chloroform	0.10	0.014	0.063	0.085	0.104	0.040
DMF[e]	0.69	Acetonitrile	0.40	0.108	0.114	0.100	0.076	0.058
HMPT[d]	1.00	Nitromethane	0.06	0.091	0.251	0.358	0.469	0.545
Dimethylsulfoxide	0.76	Dichloromethane	0.10	0.237	0.187	0.242	0.441	0.581

TABLE 2.27C Continued

Solvent A	β_A	Solvent B	β_B				x_B		
				0.15	0.35	0.50	0.65	0.85	
Dimethylsulfoxide[e]	0.76	Chloroform	0.10	0.088	0.144	0.260	0.244	0.168
Dimethylsulfoxide	0.76	Tetrachloromethane	0.10	0.308	0.325	0.385	0.534	0.582
Dimethylsulfoxide	0.76	Chlorobenzene	0.06	0.217	0.111	0.237	0.515	0.678
Dimethylsulfoxide	0.76	1,2-Dichloroethane	0.10	0.164	0.134	0.217	0.411	0.517
Dimethylsulfoxide	0.76	Acetonitrile	0.40	0.045	0.095	0.128	0.161	0.127
Dimethylsulfoxide	0.76	Nitromethane	0.06	0.108	0.244	0.336	0.424	0.509

Source: Calculated from data in Ref. 30 unless otherwise noted.

[a]RJ Sindreu, ML Moya, FS Burgos, AG Gonzalez. J Solution Chem 25:289, 1996.

[b]TM Krygowski, C Reichardt, PK Wrona, C Wyszomirska, U Zielkowska. J Chem Res (S) 1983: 116; the original data are given in terms of B_{KT}, where for the neat cosolvents $\beta = 0.002 + 0.975 B_{KT}$.

[c]MJ Kamlet, EG Kayser, ME Jones, J-LM Abboud, JW Easter, RW Taft. J Phys Chem 82:2477, 1978.

[d]P Nagy, R Herzfeld. Acta Phys Chim (Acta Univ Szeged) 31:735, 1985.

[e]PME Mancini, A Terenzani, C Adam, A Perez, LR Vottero. J Phys Org Chem 12:207, 713, 1999, from the 4-nitroaniline data.

TABLE 2.28A Deviation $\Delta\alpha$ for Mixtures of Solvent A, Water ($\alpha = 1.17$), with Cosolvents B at Various Mole Fractions x_B at the Heads of Columns

Solvent B	α_B	0.15	0.35	0.50	0.65	0.85
Methanol	1.00	−0.127	−0.128	−0.106	−0.065	−0.014
Ethanol	0.90	−0.088	−0.120	−0.073	−0.042	−0.023
i-Propanol	0.77	−0.149	−0.120	−0.063	−0.052	−0.060
Tetrahydrofuran	0.00	−0.375	−0.268	−0.081	−0.046	−0.053
1,4-Dioxane	0.00	−0.009	0.34	0.125	0.165	0.117
Acetone	0.08	−0.045	0.004	0.104	0.180	0.170
Formic acid	1.27	0.003	−0.037	−0.011	−0.007	−0.023
Acetic acid	1.12	−0.151	−0.160	−0.094	−0.052	−0.054
Pyridine	0.00	−0.404	−0.439	0.345	−0.252	−0.150
Acetonitrile	0.19	−0.023	0.075	0.184	0.272	0.304
Formamide	0.63	−0.158	−0.168	−0.130	−0.093	−0.047
Dimethylformamide	0.00	−0.330	−0.336	−0.224	−0.113	−0.028
Dimethylsulfoxide	0.00	−0.309	0.345	−0.266	−0.178	−0.081
HMPT	0.00	−0.135	−0.262	−0.272	−0.248	−0.160

Source: Calculated from data in Ref. 30 unless otherwise noted.

In *ideal associated mixtures* the activity coefficients of the monomers are all unity:

$$f_{A_1} = f_{A_1}{}^0 = f_{B_1} = f_{B_1}{}^0 = 1 \tag{2.69}$$

so that the measurable activity coefficient of component A becomes

$$f_A = x_{A_1}/x_A x_{A_1}{}^0 \tag{2.70}$$

and similarly for component B. The following discussion concerns only the simplified cases of ideal associated mixtures: (1) where component A alone associates forming dimers, the species in the mixture being A, A_2, and B, and (2) where mutual 1:1 association but no self-association takes place, the species being A, AB, and B. More complicated cases may be looked up in the works of Kehiaian et al. [40].

In case 1 the equilibrium $2A \rightleftharpoons A_2$ takes place, governed by the equilibrium constant $K_2 = x_{A_2}/x_{A_1}{}^2$. The dimer mole fraction in the mixture, x_{A_2}, is then obtained as follows:

$$x_{A_2} = x_A/(2 - x_A) - \{[4K_2 x_A(2 - x_A) + 1]^{1/2} - 1\}/2K_2(2 - x_A)^2 \tag{2.71}$$

TABLE 2.28B Deviation $\Delta\alpha$ for Mixtures of Nonaqueous Solvent A with Cosolvents B at Various Mole Fractions x_B at the Heads of Columns

Solvent A	α_A	Solvent B	α_B	x_B 0.15	0.35	0.50	0.65	0.85
Ethanol[a]	0.86	Chloroform	0.20	0.020	0.036	0.048	0.057	0.070
Ethanol[a]	0.86	Acetone	0.08	0.022	0.036	0.041	0.048	0.049
Acetone[a]	0.08	Dichloromethane	0.13	0.207	0.312	0.342	0.338	0.214
Acetone[a]	0.08	Chloroform	0.20	0.181	0.241	0.219	0.178	0.108
Dichloromethane[b]	0.13	Ethyl acetate	0.00	0.019	0.057	0.060	0.068	0.047
Dichloromethane[b]	0.13	Dimethylsulfoxide	0.00	-0.027	-0.008	0.015	0.038	0.043
Chloroform[b]	0.20	Ethyl acetate	0.00	-0.135	-0.002	0.051	0.087	0.115
Chloroform[b]	0.20	Dimethylsulfoxide	0.00	-0.068	0.062	0.068	0.082	0.124
Acetonitrile[b]	0.19	Dichloromethane	0.13	0.012	0.013	0.007	0.002	0.018
Acetonitrile[b]	0.19	Chloroform	0.20	0.020	0.002	-0.054	-0.095	-0.147
Acetonitrile[b]	0.19	Acetone	0.08	0.041	0.065	0.067	0.078	0.076
Acetonitrile[b]	0.19	2-Butanone	0.04	0.059	0.077	0.075	0.072	0.060
Acetonitrile[b]	0.19	Ethyl acetate	0.00	0.053	0.108	0.135	0.122	0.109
Acetonitrile[b]	0.19	DMF	0.00	0.000	0.008	0.017	0.022	0.033
Acetonitrile[b]	0.19	Dimethylsulfoxide	0.00	-0.044	-0.043	-0.043	-0.038	-0.011

Source: Calculated from data in the noted references.
[a]S Balakrishnan, AJ Easteal. Aust J Chem 34:933, 1981, calculated from Z, Eq. (2.63), ignoring $\Delta\pi^*$.
[b]PME Mancini, A Terenzani, C Adam, A Perez, LR Vottero. J Phys Org Chem 12:207, 1998, calculated from $E_T(30)$, Eq. (2.62), taking into account $\Delta\pi^*$ and $\Delta\beta$.

In pure component A the mole fractions of the monomer and dimer are then

$$x_{A_1}^{0} = [(4K_2 + 1)^{1/2} - 1]/2K_2 \qquad (2.72a)$$

$$x_{A_2}^{0} = 1 - [(4K_2 + 1)^{1/2} - 1]/2K_2 \qquad (2.72b)$$

The maximal possible mole fraction of the dimer is obtained if an infinitely large equilibrium constant, $K_2 = \infty$, is assumed: $x_{A_2}^{0}{}_{max} = x_A/(2 - x_A)$. In the mixture, the slope $(\partial x_{A_2}/\partial x_A)_{T,P} > 0$, except at $x_A = 0$, where it is zero. The molar excess Gibbs energy of the mixture, for any value of K_2, is

$$G^E = RTx_A \ln(x_{A_1}/x_{A_1}^{0}x_A) + RTx_B \ln(x_{B_1}/x_B) \qquad (2.73)$$

Note that although component B does not associate, in the mixture its actual mole fraction $x_{B_1} \neq x_B$, i.e., is not the nominal one. The excess molar Gibbs energy is asymmetrical with respect to the composition. So are the activity coefficients of the nominal components:

$$f_A = x_{A_1}/x_{A_1}^{0}x_A = [x_A - x_{A_2}(2 - x_A)]/x_A(1 - x_{A_2}^{0}) > 1 \qquad (2.74a)$$

$$f_B = x_{B_1}/x_B = 1 + x_{A_2} > 1 \qquad (2.74b)$$

Both activity coefficients are larger than unity and $G^E > 0$ over the entire composition range (except for the pure components). The dimer mole fraction in pure component A can be obtained from the limiting activity coefficients:

$$f_A^{\infty} = \lim f_A(x_A \to 0) = 1/(1 - x_{A_2}^{0}) \qquad (2.75a)$$

and

$$f_B^{\infty} = \lim f_B(x_A \to 1) = 1 + x_{A_2}^{0} \qquad (2.75b)$$

which can be determined by gas chromatography, depending on the volatility of one or the other component. Thus, the value of $x_{A_2}^{0}$ in the pure component A are available experimentally, so that the equilibrium constant K_2 can be obtained from Eq. (2.72b).

In spite of the ideal mixing of the species themselves that is assumed, a heat of mixing of the components A and B is measurable. It is related to the standard enthalpy of the dimerization reaction, $\Delta H_{A_2}^{0}$:

$$H^E = [x_{A_2}/(1 + x_{A_2}) - x_A x_{A_2}^{0}/(1 + x_{A_2}^{0})]\Delta H_{A_2}^{0} \qquad (2.76)$$

The term in the square brackets is negative, hence H^E and $\Delta H_{A_2}{}^0$ have opposite signs, and because the latter should be negative for the association reaction, the heat of mixing should be positive. It is asymmetrical with respect to the composition and the values of H^E range between zero and $-0.039\Delta H_{A_2}{}^0$, i.e., H^E is fairly small. The maximal value of H^E is reached for $K_2 \sim 1.1$ and $x_A \sim 0.4$ for a given standard molar enthalpy of dimerization.

In case 2 the equilibrium $A + B \rightleftharpoons AB$ takes place, governed by the equilibrium constant $K_{AB} = x_{AB}/x_{A_1}x_{B_1}$, still retaining the assumption of ideal mixing of the species present. This equilibrium reaction is symmetrical in the components, hence the excess functions are expected to be symmetrical too. The expression that is derived for the mole fraction of the adduct is

$$x_{AB}=(K_{AB}+1)/2K_{AB}x_A(1-x_A)-1-\{[(K_{AB}+1)/2K_{AB}x_A(1-x_A)-1]^2$$
$$-1\}^{1/2} \qquad (2.77)$$

For the neat components $x_{A_1} = x_A = 1$ and $x_{B_1} = x_B = 1$, since necessarily $x_{AB}=0$. The mole fraction of the adduct increases with the equilibrium constant K_{AB} and reaches the maximal values $x_{AB\ max} = x_A/(1-x_A)$ for $x_A \leq 0.5$ and $(1-x_A)/x_A$ for $x_A \geq 0.5$ as $K_{AB} \rightarrow \infty$. The excess molar Gibbs energy is

$$G^E = RT\, x_A \ln(x_{A_1}/x_A) + RT\, x_B \ln(x_{B_1}/x_B) \qquad (2.78)$$

and is necessarily negative since $x_{A_1} < x_A$ and $x_{B_1} < x_B$ in the mixture. The measurable activity coefficients of the nominal components are

$$f_A = x_{A_1}/x_A = 1 - x_{AB}(1 - x_A)/x_A < 1 \qquad (2.79a)$$

$$f_B = x_{B_1}/x_B = 1 - x_{AB}x_A/(1 - x_A) < 1 \qquad (2.79b)$$

so that the deviations from Raoult's law are negative.

In this case, again, the heat of mixing is measurable and is related to the standard enthalpy of the adduct formation, $\Delta H_{AB}{}^0$:

$$H^E = x_{AB}\Delta H_{AB}{}^0/(1 + x_{AB}) \qquad (2.80)$$

It has the same sign as the enthalpy of adduct formation and is, therefore, generally negative. An example of a nearly ideal associated mixture of this type is the acetone + choloroform system, where at 298.15 K the adduct formation equilibrium constant is $K_{AB}=0.77$ and the standard molar enthalpy of adduct formation is $\Delta H_{AB}{}^0 = -11.3$ kJ mol^{-1}[41].

It would be wrong, of course, to conclude from the experimental finding of a positive heat of mixing and activity coefficients that are larger than unity in a binary solvent mixture that ideal association of one of the components to a dimer [Eqs. (2.71) to (2.75)] takes place. Similarly, it would be wrong to conclude from the finding of a negative heat of mixing and activity coefficients smaller than unity in a binary solvent mixture that ideal association of the components to a 1:1 adduct [Eqs. (2.76) to (2.80)] take place. Other species (such as A_3, AB_2, $A_2 + B_2$) may be formed and/or the species may interact with one another nonideally. Such cases have been dealt with extensively in the literature. Here only some typical systems will be discussed.

In fact, many binary solvent mixtures split into two immiscible liquid phases (although such systems are not discussed in this book). This immiscibility requires that the molecules of the species present do not interact ideally in such systems, whether or not association takes place. If the solvent mixtures were ideally associating ones, then the association equilibrium constants, $K_{A_k B_p}$, were independent of the composition in the assumed two immiscible liquid phases of such systems. The equilibrium condition, Eq. (2.67), for the monomeric species in the two phases therefore necessarily makes these phases identical in all respects, i.e., transforms them into a single homogeneous liquid phase [39]. Solvent systems that split into two immiscible liquid phases are, therefore, non-ideal.

One way to deal with the nonidealities of the interactions in associating solvent mixtures is to assume that the species interact according to the *regular solution* model. As mentioned in Section 2.3, according to this model an average interaction parameter $\chi = g_0$ of Eq. (2.14) with all $g_{i>0} = 0$ can be written for all the species. Then, for instance, for a binary solvent mixture in which a 1:1 adduct AB is formed, Eq. (2.78) for the excess Gibbs energy of mixing is modified by inclusion of the additional term $\chi x_A(1-x_A)$. The activity coefficient f_A in Eq. (2.79a) is then multiplied by the factor $\exp(\chi x_B{}^2)$ and the factor $\exp(\chi x_A{}^2)$ multiplies f_B in Eq. (2.79b).

Consider a regular solution system in which a dimer is formed $(A + A_2 + B)$, where the mole fraction of the dimer in the pure component A is $x_{A_2}^0$. One consequence for such a system is that if the average interaction parameter $\chi > RT/(1 + x_{A_2}^0)$, then dilution of A by some B will cause *more* dimer to be formed, contrary to the intuitive expectation of what would happen on dilution. Another consequence is the dependence now of the equilibrium quotient $Q_2 = x_{A_2}/x_{A_1}^2$ on the composition, varying from $Q_2 = K_2 \exp(2\chi/RT)$ at $x_A = 0$ to $Q_2 = K_2$ at $x_A = 1$. If a 1:1 adduct is formed in a regular associated mixture, then the equilibrium quotient $Q_{AB} = x_{AB}/x_A (1-x_A)$ also varies with the composition, with a maximal value (at $x_A = 0$ or 1) and a minimal value (at $x_A = 0.5$), the ratio between these extreme values being $\chi(1+x_{AB})^2/4RT$.

A special case of association encountered in binary solvent mixtures is the formation of a *Mecke–Kempter series* [42] of oligomers of one of the components, say A, consisting of A_1, A_2, A_3, ... A_n, which is sometimes taken nominally up to $n = \infty$, together with monomeric B. Such a series is frequently invoked when component A is an alcohol or some other amphiprotic hydrogen-bonding solvent. The molecules of A are then capable of both donating and accepting hydrogen bonds and may associate to chainlike or possibly also cyclic oligomers. It is then often assumed that the addition of a further monomer to an already existing oligomer A_{i-1} takes place with the same association constant $K_{i-1,i} = K_2$ and with a constant enthalpy of association, $\Delta H_{i-1,i}^0 = \Delta H_2^0$, independent of the value of i. The interaction parameter between molecules of B and each of the species A_i are also often assumed to be $\chi_i = \chi(1+m(i-1))$. If $m = 0$ is used, then the same χ is taken for all the interactions. This is tantamount to taking B to interact with just the nearest monomer in the oligomer. Otherwise, different values result for the different oligomers. Such a series has been postulated to be formed, e.g., in mixtures of alkanols with alkanes [43], but other interpretations of the observed thermodynamic behavior are possible [44].

REFERENCES

1. A Leo, C Hansch, D Elkins. Chem Rev 71:525, 1971.
2. Y Marcus. The Properties of Solvents. Chichester, UK: Wiley, 1998.
3. DIPPR Chemical Database (http:\\dippr.byu.edu) Design Institute for Physical Property Data, Provo, UT: Am. Inst. Chem. Eng., 2000.
4. JA Riddick, WB Bunger, TK Sakano. Organic Solvents. 4th ed. New York: Wiley-Interscience, 1986.
5. L Grunberg, AH Nissan. Nature 164:799, 1940.
6. LA Robbins. Chem Eng Prog 76(10):58, 1980.
7. O Redlich, AT Kister. Ind Eng Chem 40:345, 1948.
8. JH Hildebrand, SE Wood. J Chem Phys 1:817, 1933.
9. G Scatchard. Chem Rev 8:321, 1931.
10. GM Wilson. J Am Chem Soc 86:127, 133, 1964.
11. H Renon, JM Prausnitz. AIChE J 14:135, 1968.
12. RL Scott. J Chem Phys 25:193, 1956.

13. EA Guggenheim. Proc R Soc Lond. A169:134, 1938; Mixtures. Oxford: Clarendon Press, 1952.
14. DS Abrams, JM Prausnitz. AIChE J 21:116, 1975.
15. RH Luecke. Fluid Phase Equil 14:373, 1983.
16. GM Wilson, CH Deal. Ind Eng Chem Fundam 1:20, 1962.
17. A Fredenslund, RL Jones, JM Prausnitz. AIChE J 21:1086, 1975.
18. RC Reid, JM Prausnitz, BE Poling. The Properties of Gases and Liquids. 4th ed. New York: McGraw-Hill, 1987.
19. AM Kolker, MV Kulikov, GA Krestov. Izv Vyssh Uchebn Zaved Khim Khim Tekhnol 28:11,1985.
20. P Gianni, L Lepori. J Solution Chem 25:1, 1996.
21. F Franks, DJG Ives Q Rev 20:1, 1996.
21a. D Khossravi, KA Connors. J Solution Chem 22:321, 1993.
22. J Escobedo, GA Mansoori. AIChE J 44:2324, 1998.
23. WL Marshall, EV Jones. J Inorg Nucl Chem 36:2319, 1974.
24. Y Marcus. J Solution Chem 21:1217, 1992; 25:455, 1996.
25. JC McGowan. J Appl Chem Biotechnol 28:599, 1978; 34A:38, 1984.
26. A Bondi. J Phys Chem 68:441, 1964.
27. Y Marcus. J Phys Chem 95:8886, 1991.
28. M Roses, C Rafols, E Bosch. Anal Chem 65:2294, 1993; C Rafols, E Bosch, M Roses, AG Asuero. Anal Chim Acta 302:355, 1995.
29. AG Asuero, MA Herrador, AG Gonzalez. Talanta 40:479, 1993.
30. Y Marcus. J Chem Soc Perkin Trans 2 1994:1015, 1751.
31. M Reta, R Cattana, JJ Silber. J Solution Chem 30:237, 2001.
32. C Reichardt. Solvents and Solvent Effects in Organic Chemistry. 2nd ed. Weinheim:VCH, 1988.
33. Y Marcus. Chem Soc Rev 22:409, 1993.
34. H Langhals. Angew Chem Int Ed Engl 21:724, 1982.
34a. JG Dawber, J Ward, RA Williams. J Chem Soc Faraday Trans 1 84:713, 1988.
34b. EB Tada, LP Novaki, OA El Seoud. J Phys Org Chem 13:679, 2000.
35. MJ Kamlet, J-LM Abboud, MH Abraham, RW Taft. J Org Chem 48:2877, 1983 and references therein.
36. V Bekarek. J Phys Chem 85:722, 1981.
37. I Koppel, V Palm. In Advances in Linear Free Energy Relationships. London: Plenum, 1973; I Koppel, A Paju. Org React (Tartu) 11:121, 1976.
38. EM Kosower. J Am Chem Soc 80:3253, 1958.
39. E Prigogine, J Defay. Chemical Thermodynamics. London: Longman, 1954.
40. H. Kehiaian. Bull Acad Pol Sci Ser Sci Chim 12:77, 1964; H Kehiaian, A Fajans. Bull Acad Pol Sci Ser Sci Chim 12:255, 1964; H Kehiaian. Bull Acad Pol Sci Ser Sci Chim 14:703, 1966; A Treszczanowicz, H Kehiaian. Bull Acad Pol Sci Ser Sci Chim 14:413, 1966.
41. L Sarolea-Mathot. Trans Faraday Soc 49:8, 1953.

42. H Kempter, R Mecke. Naturwissenschaften 27:583, 1939; H Kempter, R Mecke. Z Phys Chem B46:229, 1940.
43. ND Coggeshall, EL Saier. J Am Chem Soc 73:5414, 1951.
44. Y Marcus. Introduction to Liquid State Chemistry. Chichester, UK: Wiley, 1977.

3

The Structure of Solvent Mixtures

The structure of solvent mixtures, like that of neat solvents, can be determined by diffraction methods (X-ray and neutron diffraction measurements) as well as by computer simulation methods (Monte Carlo and molecular dynamics calculations). Solvent mixtures, like liquids in general, are isotropic, hence the information is generally obtained in a one-dimensional manner as the pair correlation function, $g_{ij}(r)$, describing the probability of finding a molecule i at a distance r from a molecule j. In solvent mixtures with components A and B, both $i \neq j$ (i.e., A-B correlations) and $i = j$ (i.e., A-A and B-B correlations) have to be considered. Only in rare cases can the mutual orientations of the molecules of the two components be obtained from such measurements and calculations. The information provided by $g_{ij}(r)$ pertains, therefore, to the centers of mass of the molecules or of certain atoms perceived by the diffraction methods (such as oxygen and nitrogen but generally not hydrogen nor, sometimes, carbon). Nevertheless, such information is valuable because it complements other information, obtained from thermodynamic and spectroscopic data, concerning the preference of molecules of the same or the other kind to be in the vicinity of a given molecule in the mixture.

That is, structural methods provide information concerning the occurrence or otherwise of preferential solvation.

It is instructive to compare the conclusions concerning the structures of given binary solvent mixtures obtained by the use of diffraction, computer simulation, and spectroscopic methods. Not in all cases are the interpretations of the results of the different methods consistent with each other. On the whole, however, good agreement is obtained and definite pictures of the structures of those mixtures emerge, where several methods have been employed, from a combination of the results. Furthermore, a comparison with the results for preferential solvation, obtained mainly from the use of thermodynamic methods and discussed in Chapter 4, is warranted. Again, good agreement is in general achieved, in particular as regards the phenomenon of microheterogeneity.

The binary mixtures that have been most widely studied by all these methods are aqueous methanol and aqueous acetonitrile. The former system evolves on mixing the components rather smoothly from the three-dimensional network of hydrogen bonds in water to the chainlike hydrogen bonding in methanol. For the latter there is wide acceptance of the presence of aggregates of acetonitrile and of water in a range of x_{MeCN} from, say, 0.25 ± 0.05 to 0.75 ± 0.05. However, some methods stressed the formation of water aggregates and others the formation of aggregates of acetonitrile, and complete agreement was not reached. This remains a challenge for future work.

Table 3.1 is a collection of binary solvent systems studied by diffraction, computer simulation, and spectroscopic methods as discussed in this chapter. It shows the relevant references so that the reader can compare the results. Most of these systems have also been studied in terms of preferential solvation as described in Chapter 4.

3.1 DIFFRACTION METHODS

Neutron diffraction, which has been applied to neat liquids and electrolyte solutions, has only rarely been applied to solvent mixtures. Hence the following is written in terms of X-ray diffraction. The information is obtained from the relationship of the intensity I of the monochromatic X-ray beam (of wavelength λ and intensity I_0), diffracted by the electrons in the atoms constituting the solvent molecules at an angle θ, with this angle. The atoms are thus the scattering centers, and the resulting information is made up from both *intra*molecular distances inside molecules and *inter*molecular distances between molecules, and these have

TABLE 3.1   References for the Structures of Some Binary Systems A + B Studied by Diffraction, Computer Simulation, and Spectroscopic Methods

		Diffraction methods		Computer simulations		Spectroscopic methods	
Solvent A	Solvent B	X-rays	Neutrons	MC	MD	NMR	IR
Water	Methanol	9, 17		23, 24	20–22, 25	43	
	Ethanol	7, 8, 10				43	
	n-Propanol	6, 8					
	i-Propanol	6, 8				43	
	t-Butanol	3–5, 8	16			43	43
	2-BuEtOH	8					
	Haloalkanols					45	
	THF				30		
	Dioxane	11	18				12
	DME[a]			34, 35			
	Acetone			29	28	42	42
	Pyridine			31			
	MeCN	12		36	37	46	48–50
	DMF			32			
	TMU[b]			33			
	DMSO		15		38	44	44
Methanol	Chloroform				40	47	
	Acetone				41		
	Pyridine			39			
	MeCN			41			51, 52
	Various						51
MeCN	Various						52

[a] 1,2-Dimethoxyethane.
[b] N,N,N',N'-Tetramethylurea.

to be disentangled. Generally, distances between atoms in a given molecule that are chemically bonded to each other are shorter than the sum of the van der Waals radii of the atoms and can be recognized as such. Still, only molecules with few scattering atoms (excluding hydrogen atoms) can be dealt with by X-ray diffraction of liquids in general and of solvent mixtures in particular. It is not surprising that aqueous mixtures of solvents that consist of small molecules have received major attention in this respect.

The angle-dependent intensity of the scattered radiation is generally recorded in terms of the variable $s = \lambda^{-1} 4\pi \sin(\theta/2)$ rather than directly in terms of θ. The atomic scattering intensity for isolated atoms, $a^2(s)$, depends on the angle θ. The structure factor S of a liquid depends on the number of atoms, N_Z, of atomic number Z (number of electrons in each atom) in the sample:

$$S(s) = [I(s)/I_0] / \sum N_Z Z^2 a^2(s) \tag{3.1}$$

The summation extends over all the atoms in the sample exposed to the incoming beam. At large values of s the ratio $I(s)/I_0$ tends to $\sum N_Z Z^2 a^2(s)$ so that $S(s)$ approaches unity, permitting normalization at lower values of s. The structure factor is related to the pair correlation function through the Fourier transform:

$$g(r) = 1 + (2\pi^2 \rho r)^{-1} \int_0^\infty [S(s) - 1]s(\sin sr)ds \tag{3.2}$$

where ρ is the number density of molecules in the sample. Major sources of error are the incomplete knowledge of $a^2(s)$ for small s values and the great diminution of $a^2(s)$ and $I(s)$ for large values of s, preventing the accurate determination of $I(s)$ for $s > 200$ (nm)$^{-1}$. Hence, the integral in Eq. (3.2) has to be truncated at both ends of the s scale, and spurious extrema might occur in the Fourier-transformed function, $g(r)$. Nevertheless, small-angle X-ray scattering (SAXS) does provide essential information concerning concentration fluctuations (hence, of Kirkwood–Buff integrals, see Section 4.3) and the correlation length. Small angles are characterized by $0.3 \leq s/\text{nm}^{-1} \leq 3$, from which zero angle scattering intensities are obtained by extrapolation.

For binary solvent mixtures A + B, the quantity $\sum N_Z Z^2$ in Eq. (3.1) is replaced by $(\rho_A Z_A + \rho_B Z_B)^2$, where Z_i is the number-weighted mean atomic number of solvent i. To be more precise, the zero angle scattering intensity is related to thermodynamic characteristics of the mixture [1] by

$$I(0) = k_B T V[(\rho_A Z_A + \rho_B Z_B)^2 \kappa_T + \rho_A^2 (V_A Z_A + V_B Z_B)^2 / V(\partial \mu_A / \partial N_B)] \tag{3.3}$$

where κ_T is the isothermal compressibility of the mixture, the Vs are the molar, respectively partial molar, volumes, and μ is the chemical potential. This expression is quite sensitive to the value of κ_T, which needs to be known (and rarely is, see Section 2.3) for the mixture at the temperature of the measurements. From these data the mean square fluctuations in the concentration $N\langle(\Delta x_A)^2\rangle$, in the particle number $\langle(\Delta N)^2\rangle/N$, and their correlation $\langle\Delta x_A \Delta N\rangle$ are obtained [2].

$$I(0)/N = (Z_A - Z_B)^2 N\langle(\Delta x_A)^2\rangle + (x_A Z_A + x_B Z_B)^2 \langle(\Delta N)^2\rangle/N$$
$$+ 2(x_A Z_A + x_B Z_B)^2 \langle\Delta x_A \Delta N\rangle \qquad (3.4)$$

whence the individual mean square fluctuations in the concentration, in the particle number, and in their correlation are obtainable from Eqs. (3.3) and (3.4), requiring knowledge of the thermodynamic quantities κ_T, V_A, and V_B. For instance,

$$\langle\Delta x_A \Delta N\rangle = [(N k_B T/V)(x_A Z_A + x_B Z_B)^2 - I(0)/N]/[(x_A Z_A + x_B Z_B)$$
$$- (Z_A - Z_B)/(N/V)(V_A - V_B)]^2 \qquad (3.5)$$

and

$$\langle(\Delta N)^2\rangle/N = (N/V)k_B T \kappa_T - (N/V)(V_A - V_B)\langle\Delta x_A \Delta N\rangle \qquad (3.6)$$

A further quantity obtained from the limiting scattering intensity (i.e., for $s \to 0$) is the Orenstein–Zernike correlation length ξ because at low s, $I(s)/I(0) \approx 1-\xi^2 s^2$. The Debye correlation length $L_D = 6^{1/2}\xi$, and the radius of gyration of a molecule $r_g = 3^{1/2}\xi$ can then be derived.

Small-angle X-ray scattering was applied to aqueous ethanol, 1- and 2-propanol, and 2-methyl-2-propanol by Nishikawa et al. [3–7], and the derived structural fluctuations were reported. Figure 3.1 shows the fluctuation correlation $\langle\Delta x_A \Delta N\rangle$ for these systems. The Debye correlation lengths, L_D, have maxima in dilute solutions of the alkanols: 1.2 nm at $x_{tBuOH} = 0.15$, 1.35 nm at $x_{nPrOH} = 0.15$, and 0.65 nm at $x_{iPrOH} = 0.20$. The larger these lengths, the larger are the microheterogeneous domains in the mixtures. D'Arrigo and Teixeira [8] applied small-angle neutron scattering (SANS) to the same, but dilute, aqueous alkanol mixtures (as well as to 2-butoxyethanol), "aqueous" being in these cases D_2O. The alkanol aggregation found by this method increases with increasing size of the alkyl radical, but the nature of these aggregates, whether micellelike or mixed water-alkanol complexes, could not be ascertained.

Takamuku, Yamaguchi, et al. reported investigations of wide-angle X-ray diffraction experiments pertaining to aqueous methanol (MeOH) [9], ethanol (EtOH) [10], 1,4-dioxane (Diox) [11], and acetonitrile (MeCN) [12]. In the case of aqueous methanol, the hydrogen bond distance $O\cdots H\cdots O$ decreased gradually from 0.287 ± 0.005 nm in water to 0.276 ± 0.005 nm in pure methanol and the coordination number around

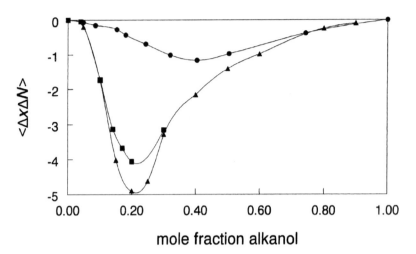

mole fraction alkanol

FIGURE 3.1 The mean correlation fluctuation, $\langle \Delta x \Delta N \rangle$, in aqueous alkanol mixtures according to Nishikawa et al. [3]; ethanol (●), n-propanol (▲), and t-butanol (■).

an oxygen atom decreased from 3.4 to 1.9. The existence of clusters with 6 to 12 methanol molecules was deduced from the data, having 24 water molecules (for clusters with 6 methanol molecules) in water-rich mixtures decreasing to 14 water molecules at $x_{MeOH} = 0.7$ and then more steeply as neat methanol was reached [9]. Similar conclusions concerning hydrogen bond distances were drawn from the data for aqueous ethanol, where clusters were found at $0.2 \leq x_{EtOH} \leq 0.8$, with water molecules bridging between chains of ethanol molecules [10]. In the case of aqueous dioxane at $x_{Diox} \leq 0.1$ the inherent water structure predominates, whereas at $x_{Diox} \geq 0.3$ that of dioxane is observed, with water molecules being hydrogen bonded to both oxygen atoms of the dioxane. In between these limits, clusters involving one or two dioxane and several water molecules are formed [11]. Neat acetonitrile has an alternating antiparallel–parallel (i.e., zigzag) orientation of the dipoles of the molecules of this very polar solvent. When water was added to acetonitrile, water molecules were dispersed in this structure and also interacted by dipole–dipole interactions in an antiparallel manner. However, when the acetonitrile content diminished to $x_{MeCN} \leq 0.8$ clustering of the acetonitrile molecules was observed, with water molecules surrounding the clusters but not being incorporated in them. When the water content was increased further, water clusters with enhanced hydrogen bonding were observed in the range $0.2 \leq x_{MeCN} \leq 0.6$, leading to microheterogeneity [12] (see Section 4.3).

In an earlier study, Radnai et al. [13,14] investigated two mixtures of the strongly hydrogen bond donating 2,2,2-trifluoroethanol (TFE) and the strong hydrogen bond accepting dimethylsulfoxide (DMSO) with compositions given by $x_{TFE} = 0.50$ and 0.67. It was concluded that besides the inherent structures of the neat solvents, the mixtures contained hydrogen-bonded linear complexes DMSO$\cdots$TFE and TFE$\cdots$DMSO$\cdots$TFE. Because clusters with the structures of the neat components were preserved, microheterogeneity was said to exist in these mixtures [13]. In a study of a mixture of acetonitrile (MeCN) and N,N-dimethylformamide (DMF) at $x_{MeCN} = 0.5$ it was found that each DMF molecule has on the average 1.8 MeCN nearest neighbors and vice versa: each MeCN has 1.8 DMF nearest neighbors, with antiparallel configurations of their dipoles [14].

In all the studies of solvent mixtures by wide-angle X-ray diffraction, it was necessary to assume models for the interactions and mutual configurations of the molecules and try to fit the observed structure factors $S(s)$ or pair correlation functions $g(r)$ to these models. The one yielding the smallest mean square deviations of the calculated from the experimental curves was deemed the most nearly correct representation of the system. There was no way of directly disentangling the observed peaks into those due to *intra*molecular and *inter*molecular distances.

Neutron diffraction with hydrogen–deuterium isotope substitution (or, in principle, isotope substitution of other elements) does permit structural conclusions to be drawn without resorting to a model because it produces partial (atom-specific) structure factors and pair correlation functions for the substituted atoms. Soper et al. [15–17] used this technique to study the structure of binary mixtures. Thus, in aqueous dimethylsulfoxide [15] they showed that the water structure is not strongly affected by the presence of the DMSO. However, a substantial fraction of the water molecules were hydrogen bonded not to other water molecules but to DMSO ones. In aqueous methanol this method substantially confirmed the results obtained by X-ray diffraction [17]. A similar technique was applied to water-rich aqueous dioxane, $x_{Diox} = 0.167$ [18]. The water hydrogen bond network was similar in such mixtures to that in pure water, the dioxane molecules being incorporated into this structure without distorting it substantially. The interaction in dilute aqueous t-butanol ($x_{BuOH} = 0.06$) was found by neutron diffraction to take place by direct methyl–methyl contacts rather than by clathrate-like hydrate cages [16], contrary to conclusions from SAXS studies [3,4].

3.2 COMPUTER SIMULATION METHODS

An alternative to experimental studies of liquid structure by means of X-ray or neutron diffraction methods is its investigation by means of computer simulations. In the present context of binary solvent mixtures, a fairly large number (typically about 250 to 1000) of molecules of the two components at various ratios are confined at random positions and orientations in a given volume. In the case of the molecular dynamics (MD) method, they are also assigned random vectorial velocities and allowed to move and collide. In the case of the Monte Carlo (MC) method, a molecule selected at random is moved in a random direction for a random distance. The key quantities that govern the behavior of the system and its tendency to reach eventually a state of equilibrium are the forces between the molecules (or more specifically, the atoms constituting them), expressed as molecular potentials.

In MD simulations, Newton's laws of motion are applied to the molecules in the liquid that are moving relative to each other, considering both the potential and kinetic energies. New positions, orientations, and velocities are computed at regular intervals (of fsec to psec length) for as long as the facilities of the computer permit, typically a few nsec. In MC simulations, the configurational potential energy of the system is computed after each step of a molecule being moved at random. The step is admitted if the potential energy is essentially lowered, and this is repeated for a few million steps or so, until random fluctuations in the potential energy of the system rather than a downward trend are reached.

Both methods yield the pair correlation functions $g_{ij}(r)$ between molecules (atoms) in the solvent mixture and thermodynamic quantities, derived from the configurational potential energy of the system. These depend on the number density of molecules and either the temperature or the total energy selected for the system, i.e., the *N-P-T* or the *N-V-E* statistical ensembles. The MD method yields in addition dynamic information, e.g., self-diffusion coefficients. Because *intra*molecular relative motions of groups of atoms in multiatomic molecules also have to be taken into account, additional information on vibrational modes of bonded atoms and groups is thereby gained, to be compared with experimental data.

In dense fluids such as the solvent mixtures treated here, the pair correlation function is a function of the potential of mean force $U_{ij}(r)$ pertaining to the selected kinds of molecules, i and j:

$$g_{ij}(r) = \exp[-U_{ij}(r)/k_B T] \tag{3.7}$$

The problem now is to relate $U_{ij}(r)$ to the assumed pair potentials between molecules (or as mentioned earlier, more specifically, the atoms constituting them) $u_{ij}(r)$. A common form of the potential energy for pair interaction is the combination of coulombic interactions with the Lennard–Jones expression:

$$u_{ij}(r) = \sum q_i q_j/r + 4\varepsilon_{ij}[(\sigma_{ij}/r)^{12} - (\sigma_{ij}/r)^6] \tag{3.8}$$

where the qs are the partial charges on the atoms, σ is the molecular diameter, and ε_{ij} is a characteristic energy. For the unlike molecules (atoms) the Lorentz–Berthelot rules are generally assumed: $\sigma_{AB} = (\sigma_A + \sigma_B)/2$ and $\varepsilon_{AB} = (\varepsilon_A \varepsilon_B)^{1/2}$. The coulombic interactions are of a long-range nature and a proper summation method has to be used. A major difficulty in the application of both simulation methods, MD and MC, is, therefore, the selection of the proper pair potentials. It is necessary, however, to take into account not only direct pairwise interactions between a given molecule and all the other molecules in the system but also triplet interactions, and this complicates the problem enormously. Furthermore, because the molecules of the solvents discussed are multiatomic, it is necessary to consider direct interactions between atoms rather than molecules because of the small range of the exerted forces when uncharged particles are involved (i.e., not ions). Sometimes it is expedient to consider interactions between so-called pseudoatoms (e.g., $-CH2-$ or $-CH_3$ groups rather than the individual C and H atoms) with other atoms, such as O, or groups, such as $-OH$. Such segments can be treated, as a good approximation, as if they were single atoms located at their center of mass but possessing point dipoles and/or partial charges on different atoms.

In view of this, it is not surprising that most of the computer simulation studies have dealt with solvent mixtures in which one of the components was water or methanol, which have few atoms and small groups ($-CH_3$ and $-OH$). The pair potentials are generally taken from those established in the neat solvents, with additional ones selected when necessary (if atoms or groups not common to the two components are present) to give agreement with experimental data. Of the studies that involved water or methanol, the majority, again, concerned binary aqueous methanol mixtures.

The potentials and models that have been used for the MD simulations of binary aqueous mixtures with cosolvents have been briefly reviewed by Hawlicka [19]. Some of the relevant parameters of Eq. (3.8) are shown in Table 3.2, but others have also been employed .

Molecular dynamics calculations by Palinkas et al. [20–22], following earlier MC simulations, showed that in aqueous methanol water molecules lose and methanol molecules gain in the numbers of nearest neighbors compared with their environments in the neat solvents. The highest number of molecules with the maximal number of hydrogen bonds was observed at $x_{MeOH} = 0.25$, where the excess thermodynamic functions have extrema. In water-rich mixtures methanol molecules build into the hydrogen-bonded network of the water structure and increase the number of defects in it, whereas in methanol-rich mixtures the water molecules enter the linear hydrogen-bonded chains, increasing their branching. The peak in $g_{O-O}(r)$ found at 0.285 ± 0.002 nm is in agreement with that found by X-ray diffraction. This distance and the $g_{O-H}(r)$ peak at 0.190 nm were confirmed by subsequent MC [23,24] and MD [25]

TABLE 3.2 Examples of the Parameters for the Calculation of Pairwise Potentials for Aqueous Cosolvent Mixtures [19] (C_M Is the Pseudoatom CH_3)

Paramter	Water	Methanol	Acetonitrile	Dimethylsulfoxide
Bond length, nm	O–H 0.0957	O–H 0.0945	C–N 0.1157	S–O 0.1490
		C_M–O 0.1425	C_M–C 0.1425	C_M–S 0.1800
Angle, deg	HOH 104.52	C_MOH 108.53	C_MCN 180.00	C_MSO 107.2
				C_MSC_M 97.7
Charge, q^a	O–0.834	O–0.728	N–0.43	O–0.459
	H 0.417	H 0.431	C 0.28	S 0.139
		C_M 0.297	C_M 0.15	C_M 0.160
ε^b, kJ/mol	0.638	O 0.0737	N 0.711	O 0.276
		H 0	C 0.628	S 0.844
		C_M 0.739	C_M 0.866	C_M 0.339
σ^b, nm	0.315	O 0.308	N 0.320	O 0.294
		H 0	C 0.365	S 0.356
		C_M 0.386	C_M 0.378	C_M 0.360

[a] In proton charge units.
[b] Depth (ε) and width (σ) of Lennard–Jones potential energy well.

simulations, but it was demonstrated that the detailed results are quite sensitive to the selected potentials. The unidimensional pair correlation function $g_{O-O}(r)$ cannot provide information on the three-dimensional structure of the liquid mixtures, but more recent MD calculations by Laaksonen et al. [26] permitted spatial distribution functions to be obtained that confirmed on the whole the previous findings. These and similar findings for aqueous methanol have been reviewed by Hawlicka [19], who also reviewed results of Tanaka et al. [27] for dilute aqueous t-butanol ($x_{tBuOH} = 0.03$), where the water hydrogen-bonded structure is most enhanced.

Aqueous acetone mixtures were studied by Ferrario et al. [28] by the MD method, the $g_{O-O}(r)$ peak at 0.28 nm being strongly enhanced compared with that of water, the main enhancement being due to stronger water–water interactions rather than to water–acetone ones. The number of hydrogen bonds accepted by an acetone molecule decreases nearly linearly from about one to zero as the acetone concentration increases from dilute to neat acetone, but acetone disrupts the water structure considerably. These conclusions were corroborated by an MC simulation study by Freitas et al. [29], who have also studied aqueous tetrahydrofuran (THF) [30], pyridine [31], N,N-dimethylformamide (DMF) [32], and N,N,N',N'-tetramethylurea (TMU) [33]. In THF-rich and in aqueous pyridine mixtures the water–water coordination numbers were found to be enhanced. In aqueous DMF and TMU, strong hydrogen bonding between the water and the carbonyl oxygen atom of these amides was the predominant feature. In dilute aqueous TMU, the influence of the TMU on the hydrogen bonding of the water is negligible. Molecular dynamics simulation studies by Bedrov et al. [34,35] of 1,2-dimethoxyethane (and most recently also 1,2-dimethoxypropane) pointed to clustering of water molecules, i.e., enhanced hydrogen-bonded water structure in the ether-rich mixtures, as in aqueous THF [30]. In very dilute aqueous acetonitrile, $x_{MeCN} = 0.004$, MC calculations by Dunn and Nagy [36] yielded coordination numbers of 1.4 water molecules hydrogen bonded to the nitrogen atom and 19 water molecules in the first hydration shell of the methyl group. Also, MD simulations at $x_{MeCN} = 0.12$ [37] showed hydrogen bonding between the water hydrogen atom and the nitrogen atom at a distance N–H of 0.185 nm with a coordination number of 1.0 for the nitrogen atom. This number decreases as the acetonitrile concentration increases. Finally, in aqueous DMSO, MD simulations by Luzar and Chandler [38] showed that a DMSO molecule forms on the average two hydrogen bonds with water molecules, which have a considerably longer lifetime than those in pure water and a somewhat longer one than those of water–water interactions in the mixtures.

Fewer completely nonaqueous binary solvent mixtures have been studied by computer simulation methods, and they typically involve methanol as one of the components. Mixtures of methanol and pyridine were studied by Sinoti et al. [39] using MC simulations, with hydrogen bonded heterodimers being found, but pyridine, in an equimolar mixture, has only a small influence on the methanol–methanol association, being similar to that in neat methanol. Mixtures of methanol and chloroform, studied by MD [40], showed preferential solvation of the hydrophilic parts of the molecules by methanol and the hydrophobic ones by chloroform, resulting in microheterogeneity. Mixtures of methanol with acetone or with acetonitrile and mixtures of acetone with acetonitrile have been studied by Venables and Schmuttenmaer [41] using MD. The methanol chains present in neat methanol are shortened by the cosolvents due to hydrogen bonding of methanol to molecules of the latter rather than to other methanol molecules. Still, methanol molecules tend to remain in chains, albeit short ones, even in mixtures dilute in methanol.

3.3 SPECTROSCOPIC METHODS

Spectroscopic methods for the investigation of the structure of binary solvent mixtures may involve a probe, whose molecules show spectral changes that are sensitive to the environment of the probe. Such methods are dealt with in a subsequent section 5.2.2. Here the application of spectroscopy to the binary solvent mixture itself, in the absence of an additional solute, is discussed. A great variety of spectroscopic methods have found use for this purpose, foremost of which are nuclear magnetic resonance (NMR) and infrared (IR) vibrational spectroscopy, each having its advantages and drawbacks. Furthermore, many of the studies pertain to dilute solutions of organic materials, which constitute liquid solvents when neat, in some solvent, rather than to binary solvent mixtures over a wide composition range. These studies are for the most part outside the scope of the present book. This still leaves a great many studies that cannot be reviewed here, and only the salient features of the methods are presented.

Proton NMR studies of binary solvent mixtures, in particular aqueous cosolvent mixtures, have generally been conducted by measuring the chemical shift of the water protons, using the proton signal of C–H bonds in the cosolvent as an internal reference. The chemical shift of water protons in neat water, $\delta(^1H,OH_2)$, relative to the commonly used tetramethylsilane (TMS) as an external reference, is 4.8 ppm at room temperature. In binary aqueous mixtures $\delta(^1H,OH_2)$ is generally lower, and from this lowering conclusions on the state of polarization of the water, hence on its hydrogen bonding, are drawn. Recent investigations, however, have shown that the C–H bonds are themselves polarized in the mixtures, depending on the composition, hence should not be employed as an internal reference. Furthermore, the composition dependence of the correction for the volume magnetic susceptibility is not trivial with respect to the changes in the chemical shifts of the water protons. There-fore an external double reference has been suggested by Mizuno et al. [42–45], consisting of a capillary with an attached small sphere at its bottom, both containing TMS, and placed concentrically in the NMR tube con-taining the investigated solvent mixture. The chemical shift, $\Delta\delta_{TMS}$, of the TMS signals between the sphere and capillary permits correction for the volume magnetic susceptibility of the solvent mixture as well as the determination of chemical shifts referred to TMS:

$$\delta_{cor} = \delta_{obs} - (4\pi/3k)\Delta\delta_{TMS} \tag{3.9}$$

where $k = 4.09 \pm 0.06$ was obtained from the volume magnetic suscept-ibilities of neat solvents.

In this manner it was found [43], for instance, that in very water-rich mixtures of t-butanol, $x_{tBuOH} \sim 0.05$, the methyl proton chemical shift, $\delta(^1H,CH_3)$, has a maximum of ~ 0.05 ppm, definitely outside the experi-mental error, above the constant values reached when t-butanol is added to water at $x_{tBuOH} > 0.16$. A similar nonconstancy of $\delta(^1H,CH_3)$ was found for aqueous acetone [42], with a maximum at $x_{Me_2CO} = 0.04$, and a larger deviation from constancy was found for $\delta(^1H,CH_3)$ in aqueous dimethyl-sulfoxide [44]. These findings, as well as corresponding results for the IR spectra for the stretching frequencies of the C–H bonds (see later), point to an appreciable polarization of these bonds by the surrounding water molecules. In other words, there is evidence for hydrogen bonding of the methyl protons to the oxygen atoms of the structured water cage around the methyl group [42–44].

The chemical shifts of the oxygen-bonded protons, $\delta(^1H,OH)$, shed further light on the structures of these mixtures and those of the aqueous lighter alkanols. The $\delta(^1H,OH_2)$ signals from water are distinct from those of the hydroxyl groups of the alkanols, $\delta(^1H,ROH)$, at $x_{ROH} > 0.1$ (0.4 for methanol). The composition dependence of the former shows a gradually increasing negative slope in the series of aqueous methanol < ethanol <

n-propanol $<$ i-propanol $<$ t-butanol [43]. In very water-rich mixtures the (time-averaged) $\delta(^1H,OH)$ shows a maximum, indicating that the polarization of the water is larger than in neat water. In other words, the hydrogen-bonded water structure is enhanced in the presence of low alkanol concentrations. The (distinct) alcohol hydroxyl $\delta(^1H,OH)$ values for ethanol and the two propanols have a shallow minimum near $x_{ROH} \sim 0.6$, indicating water–alcohol association by hydrogen bonding, and the upturn in alcohol-rich mixtures points to self-association of these alcohols. This is not the case for t-butanol, where the methyl groups provide steric hindrance to extensive self-association. In all these systems, as well as in aqueous DMSO, the $\delta(^1H,OH)$ signal in the water-rich region is larger than in pure water (e.g., 4.95 ppm at $x_{DMSO} > 0.8$ compared with 4.80 ppm in neat water at 296.5 K, referred to TMS), signifying larger polarization of the water molecules, more extensive hydrogen bonding, and enhanced water structure [42–45].

Significant conclusions concerning the structure of binary solvent mixtures can still be obtained even if the precaution of using an external double reference, necessary for detecting small differences in the chemical shifts, is not used and the internal reference of C–H protons is employed instead. An illustration of this was provided by Easteal's studies on aqueous acetonitrile (MeCN) mixtures by both proton and ^{13}C NMR [46]. The $\delta(^1H,OH)$ shift in pure water was 2.62 ppm (referred to the methyl C–H proton at 308 K), and as MeCN was added it decreased initially with the 3/2 power of the cosolvent mole fraction up to $x_{MeCN} \sim 0.05$. At $x_{MeCN} > 0.20$ the change of the slope of $\delta(^1H,OH)$ versus x_{MeCN} is consistent with microheterogeneity of the mixture. This means that water molecules form regions where the water structure is retained, and in between such regions the acetonitrile molecules are dispersed. This microheterogenous composition ranges up to $x_{MeCN} \sim 0.70$, beyond which the slope changes again and indicates progressive formation of the dipole–dipole structure of MeCN in which water molecules are dispersed. The values of $\delta(^1H,OH)$ reached 0.16 ppm in infinitely dilute water in MeCN. This decrease in $\delta(^1H,OH)$ relative to 2.62 ppm in liquid water is similar to that observed when the latter is vaporized (at 453 K), indicating the state of the water at low concentrations in MeCN. The picture arising from the $\delta(^{13}C,CN)$ data is in general agreement with that from the $\delta(^1H,OH)$ data [46].

In an attempt to be rid of NMR-specific quantities (such as chemical shifts in Hz or ppm), the use of the relative shift in a mixture was recommended by Tkadlekova et al. [47]. This is the ratio of the difference of the observed chemical shift, δ, and that at infinite dilution in the cosolvent, δ^∞, to the difference between that in the neat solvent, δ°, and δ^∞. The resulting quantity:

$$\eta_{rel} = (\delta - \delta^\infty)/(\delta^\circ - \delta^\infty) \qquad (3.10)$$

in a mixture involving a hydrogen-bonding solvent (e.g., methanol) is related to the fraction of non–hydrogen-bonded protons f_{nhb} as $\eta_{rel} = (1 - f_{nhb})/(1 - f^\circ_{nhb})$. Therefore, if the value of the denominator is known, the NMR data provide information on this fraction for the entire composition of the mixture. This concept was applied to mixtures of methanol and chloroform and this fraction was compared with the expectation from a model, specifying the formation of methanol tetramers and 1:1 methanol–chloroform adducts, which was consistent with thermodynamic data on this system [47].

The application of infrared spectroscopy to binary aqueous cosolvent mixtures with respect to the O–H vibrations of the water suffers from the very intense absorption and strong overlapping of the stretching bands at the 3000 to 3600 cm^{-1} region, which prescribes the use of very thin samples. One way to overcome this is the use of attenuated total reflectance measurements that sample only the thin surface layer of the liquid. It must then be assumed that there is no excess surface concentration of the cosolvent, although it has a lower surface tension than water (see Section 2.3). A further way to overcome this difficulty is to measure the absorption spectrum of the O–D band of dilute HOD in the H_2O + cosolvent mixtures.

As illustrations, these methods concerning O–H vibrations were applied to aqueous acetonitrile mixtures, where complementary data can, of course, also be obtained from the C–N vibrations of the cosolvent. Wakisaki et al. [48] used thin, 0.025 mm, layers of the solvent mixtures and reported in a diagram the wavenumbers of what they called the v_3 (i.e., asymmetric stretching) O–H vibrations of the water over the entire composition range. They did not specify how they deconvoluted this particular vibration from the observed band or the precision of the plotted points. The diagram showed three distinct steps in the shifts of v_3 to higher wavenumbers as the acetonitrile content increased: from ~ 3430 cm^{-1} in pure water to ~ 3670 cm^{-1} for very dilute water in MeCN. These steps were interpreted as signifying the presence of several types of structures around water molecules. At $x_{MeCN} < 0.2$ the structure of pure water is maintained, but beyond this value water clusters surrounded by acetonitrile molecules appear, to disintegrate when x_{MeCN} exceeds 0.8, where the dipole–dipole structure of the acetonitrile is manifested and water molecules are incorporated into such structures [48].

Jamroz et al. [49] used dilute HOD solutions ($\sim 8\%$ of the total water) in the mixtures and found that at $x_{MeCN} \geq 0.96$ only monomeric HOD is present, with an O–D stretching vibration at 2631 cm^{-1}. This is hydrogen bonded to acetonitrile molecules (or at most to a single other water molecule), but at lower x_{MeCN} self-associated water oligomers start to be seen, the deconvoluted peak for (HOD)$_n$ moving toward 2502 cm^{-1}. Bertie and Lan [50] used attenuated total reflectance spectra to obtain the real and imaginary refractive index spectra, from which the integrated absorption intensities of the bands were obtained. In the case of the O–H stretching vibration band, a small correction for the integrated absorption of the C–H vibration at similar wavenumbers was applied. Three types of O–H groups were discerned: non–hydrogen-bonded (subscript OH), hydrogen-bonded to acetonitrile (subscript OHN), and hydrogen-bonded to water (subscript OHO). The total absorption intensity of the O–H band per mole of mixture was

$$I_{OH} = 2x_{H_2O}[f'_{OHN}I'_{OHN} + f'_{OHO}I'_{OHO} + f'_{OH}I'_{OH}] \qquad (3.11)$$

where f' is the fraction of O–H groups bonded as specified and I' is the absorption intensity per mole of the specified O–H group. The intensity of this band in dilute water vapor, where only non–hydrogen-bonded molecules exist, is 7% of that in liquid water, hence $I'_{OH} \approx 0.07 I'_{OHO}$. The fraction of hydrogen-bonded MeCN molecules is given by $x_{MeCN} f'_{CNH}$ (see later). Per mole of O–H groups in water this becomes $x_{MeCN} f'_{MeCN}/2x_{H_2O}$, from which expression the fraction of water molecules so bonded, f'_{OHN}, is obtained. The quantity $f'_{OHN} = 0.45 \pm 0.10$ was obtained from the limiting slope of a plot of $x_{MeCN} f'_{MeCN}$ versus x_{MeCN} near infinite dilution of water in acetonitrile. These data provided all that was necessary to evaluate Eq. (3.11) over the entire composition of the mixtures, and a diagram showing the fractions of these types as a function of the overall composition was presented. The fraction of non–hydrogen-bonded O–H groups decreases slightly from pure water to $x_{MeCN} = 0.10$ and then rises steadily at higher MeCN concentrations. The fraction of O–H groups bonded to acetonitrile increases steadily, whereas that bonded to water molecules diminishes steadily with increasing acetonitrile content. These relative fractions are shown in Fig. 3.2.

The information obtained from the infrared spectra of the water bands in the aqueous acetonitrile mixtures is supplemented by that obtained from the v_2 C–N stretching vibrations. In aqueous deuterated

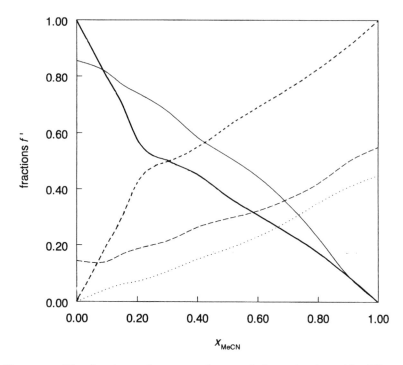

FIGURE 3.2 The fractions of water and acetonitrile molecules with different hydrogen bonding characteristics: Bold curves: (——) f'_{CNH}, (– – – – –) f'_{CN}; thin curves: (——) f'_{OHO}, (– – –) f'_{OHN}, (......) f'_{OH}, according to Bertie and Lan [50].

acetonitrile, CD_3CN, the band at 2263 cm^{-1} is due to free CN groups whereas the neighboring band at 2267 cm^{-1}, separated from the former by deconvolution of the observed asymmetric band, is due to hydrogen-bonded CN groups [49]. The attenuated total reflectance method [50] yielded integrated absorption intensity values for the C–N stretching band in aqueous CH_3CN that could be related to the relative amounts of hydrogen-bonded $C\equiv N\cdots H–O$ and non–hydrogen-bonded $C\equiv N$:

$$I_{CN} = x_{MeCN}[f'_{CN}I'_{CN} + f'_{CNH}I'_{CNH}] \tag{3.12}$$

The quantity I'_{CN} was obtained from pure acetonitrile, where the C–N group is not hydrogen bonded, and I'_{CNH} was obtained from the limiting slope at infinite dilution of the acetonitrile. The fractions in Eq. (3.12) were in general agreement with those obtained from deuterated acetoni-

TABLE 3.3 Solvent-Induced Frequency Shifts of the C–N Stretching Vibration v_2 of Acetonitrile [52] and the Gutmann Donor Number DN and the Mayer and Gutmann Acceptor Number AN of the Solvents

Solvent	DN	AN	$\Delta v_2/\mathrm{cm}^{-1}$
Hexamethyl phosphoric triamide	38.8	9.8	−5.5
Dimethylsulfoxide	29.8	19.3	−4.5
N, N-Dimethylacetamide	27.8	13.6	3.5
N, N-Dimethylformamide	26.6	16.0	−2.5
N-Methylformamide	27.0	32.1	−1.5
Propylene carbonate	15.1	18.3	−1.0
Benzonitrile	11.9	15.5	−1.0
Nitrobenzene	4.4	14.8	−1.0
Acetone	17.0	12.5	−0.5
Benzene	0	8.2	0
1,2-Dichloroethane	0	16.7	0
Acetonitrile	14.1	18.9	0
Nitromethane	2.7	20.5	0
Diethyl ether	19.2	3.9	0.5
Formamide	24	39.8	0.5
Dichloromethane	1.0	20.4	1.0
Tetrachloromethane	0	8.6	1.5
Chloroform	4.0	23.1	2.0
Hexane	0	0	2.5
Methanol	30.0	41.5	5.5
Benzyl alcohol	23.0	36.8	5.5
t-Butanol	38.0	27.1	6.0
i-Propanol	36.0	33.8	6.0
Ethanol	32.0	37.1	6.5
Acetic acid	20.0	52.9	12.0
Trifluoroacetic acid		105.3	25.5

trile [49]. These relative fractions are also shown in Fig. 3.2. The breaks in the slopes of the fractions f'_{CN} and f'_{CNH} near $x_{MeCN} \sim 0.2$ correspond to the onset of the microheterogeneity arrived at from other kinds of data. The shallow minimum in f_{OH} denotes enhanced self hydrogen bonding of the water molecules that corresponds to the exothermic mixing observed in this region.

A still further method for the investigation by infrared spectroscopy of binary solvent mixtures involving protic solvents with O–H groups is the use of overtones of the fundamental vibrations, e.g., $2v_{OH}$, applied to mixtures of methanol with cosolvents by Robinson and Symons [51]. When a basic aprotic cosolvent is mixed with methanol it breaks the hydrogen-bonded chains of the methanol, accepting itself a hydrogen bond from the fragment at one end of the chain and causing an extra "free lone pair" to appear in the terminal methanol molecule of the other fragment. Thus, two species are formed that are spectroscopically distinguishable from bulk methanol. Deconvolution of the overtone band near 7000 cm^{-1} yielded a band at 6725 cm^{-1} assigned to $2\nu_{OH}$ in the free lone pair terminal methanol molecules. This band grows in the spectrum as the basic cosolvent is added, the slope of this growth in dilute mixtures depending on the nature of this cosolvent and describing roughly the number of methanol molecules bonded to it. These numbers are ≤ 1 for (deuterated) acetonitrile; 1 for dimethylacetamide, tetrahydrofuran, and dimethylsulfoxide; between 1 and 2 for hexamethyl phosphoric triamide; and 2 for triethylphosphine oxide [51]. The numbers increase with the basicity and number of "free electron pairs" of the molecules of the cosolvent.

A final illustration of the ability of infrared spectroscopy to provide information on the interactions and structure of binary solvent mixtures is the solvent-induced frequency shift of the ν_2 C–N stretching vibration in various solvents studied by Fawcett et al. [52]. The shifts $\Delta\nu_2$ of the resolved high-frequency component of the asymmetrical band with respect to that in pure acetonitrile ($\nu_2 = 2253.5$ cm^{-1}) correlate well with the Lewis acidities and basicities of the cosolvents: a blue shift (to higher frequencies) is noted for the acidic solvents and a red shift for the basic ones. The former interact by hydrogen bonding to the lone pair of the C–N group and the latter by dipole interactions with the positive end of the large dipole of acetonitrile residing on the methyl group ($\mu = 3.92$ D, 1 D $= 3.33564 \times 10^{-30}$ C·m). Although the correlation with the Gutmann donor number DN, measuring the Lewis basicity, and the Mayer–Gutmann acceptor number, AN, measuring the Lewis acidity, is not perfect (Table 3.3), the trends are obvious.

REFERENCES

1. MCA Donkersloot. J Solution Chem 8:293, 1979.
2. H Hayashi, K Nishikawa, T Iijima. J Appl Crystallogr 23:134, 1990.
3. K Nishikawa, Y Kodera, T Iijima. J Phys Chem 91:3694, 1987.
4. K Nishikawa, H Hayashi, T Iijima. J Phys Chem 93:6559, 1989.
5. K Nishikawa, T Iijima, J Phys Chem 94:6227, 1990.

6. H Hayashi, K Nishikawa, T Iijima. J Phys Chem 94:8334, 1990.
7. K Nishikawa, T Iijima. J Phys Chem 97:10824, 1993.
8. G D'Arrigo, J Teixeira. J Chem Soc Faraday Trans 86:1503, 1990.
9. T Takamuku, T Yamaguchi, M Asato, N Nishi. Z Naturforsch A55:513, 2000.
10. M Matsumoto, N Nishi, T Furusawa, M Saita, T Takamuku, M Yamagami, T Yamaguchi. Bull Chem Soc Jpn 68:1775, 1995.
11. T Takamuku, M Tabata, M Kumamoto, A Yamaguchi. J Nishimoto, H Wakita, T Yamaguchi. J Phys Chem B102:8880, 1998.
12. T Takamuku, A Yamaguchi, M Tabata, N Nishi, K Yoshida, H Wakita, T Yamaguchi. J Mol Liq 83:163, 1999.
13. T Radnai, S Ishiguro, H Ohtaki. Chem Phys Lett 159:532, 1980.
14. T Radnai, S Itoh, H Ohtaki. Bull Chem Soc Jpn 61:3845, 1988.
15. A K Soper, A Luzar. J Chem Phys 97:1320, 1992.
16. DT Bowron, JI Finney, AK Soper. J Phys Chem B102:3551, 1998.
17. AK Soper, JL Finney. Phys Rev Lett 71:4346, 1993.
18. I Bako, G Palinkas, JC Dore, HE Fischer. Chem Phys Lett 303:315, 1999.
19. E Hawlicka. Pol J Chem 70:821, 1996.
20. G Palinkas, E Hawlicka, K Heinzinger. Chem Phys 158:65, 1991.
21. G Palinkas, I Bako, Z Naturforsch. 46A:95, 1991.
22. G Palinkas, I Bako, K Heinzinger, P Bopp. Mol Phys 73:897, 1991.
23. LCG Freitas. J Mol Struct. (Theochem) 282:151, 1993.
24. CA Koh, H Tanaka, JM Walsh, KE Gubbins, JA Zollweg. Fluid Phase Equil 83:51, 1993.
25. I Bako, G Palinkas, K Heinzinger. Z Naturforsch 49A:967, 1994.
26. A Laaksonen, PG Kusalik, IM Svishchev. J Phys Chem A101:5910, 1997.
27. H Tanaka, K Nakanishi, H Touhara. J Chem Phys 81:4065, 1984.
28. M Ferrario, M Haughney, IR McDonald, ML Klein. J Chem Phys 93:5156, 1990.
29. LCG Freitas, JMM Cordeiro, FLL Garbujo. J Mol Liq 79:1, 1999.
30. LCG Freitas, JMM Cordeiro. J Mol Struct (Theochem) 335:189, 1995.
31. ALL Sinoti, JRDS Politi, LCG Freitas. J Braz Chem Soc 7:133, 1996.
32. JMM Cordeiro, LCG Freitas. Z Naturforsch 54A:110, 1999.
33. P Belletato, LCG Freitas, EPG Areas, PS Santos. Phys Chem Chem Phys 1:4769, 1999.
34. D Bedrov, O Borodin, GD Smith. J Phys Chem B102:5683, 9565, 1998.
35. D Bedrov, GD Smith. J Phys Chem B103:3791, 10001, 1999.
36. WJ Dunn III, PI Nagy. J Phys Chem 94:2099, 1990.
37. H Kovacs, A Laskonen. J Am Chem Soc 113:5596, 1991.
38. A Luzar, D Chandler. J Chem Phys 98:8160, 1993.
39. ALL Sinoti. JRDS Politi, LCG Freitas. J Mol Struct (Theochem) 366:249, 1996.
40. R Grátias, H Kessler. J Phys Chem B102:2027, 1998.
41. DS Venables, CA Schmuttenmaer. J Chem Phys 113:3249, 2000.

42. K Mizuno, T Ochi, Y Shindo. J Chem Phys 109:9502, 1998.
43. K Mizuno, Y Kimura, H Morichika, Y Nishimura, S Shimada, S Maeda, S Imafuji, T Ochi. J Mol Liq 85:139, 2000.
44. K Mizuno, S Imafuji, T Ochi, T Ohta, S Maeda. J Phys Chem B104:11001, 2000.
45. K Mizuno, K Oda, S Maeda, Y Shindo, A Okamura. J Phys Chem 99:3056, 1995.
46. AJ Easteal. Aust J Chem 32:1379, 1979.
47. M Tkadlekova, V Dohnal, M Costas. Phys Chem Chem Phys 1:1479, 1999.
48. A Wakisaka, Y Shimizu, N Nishi, K Tokumaru, H Sakuragi. J Chem Soc Faraday Trans 88:1129, 1992.
49. D Jamroz, J Strangret, J Lindgren. J Am Chem Soc 115:6165, 1993.
50. JE Bertie, Z Lan. J Phys Chem B101:4111, 1997.
51. HL Robinson, MCR Symons. J Chem Soc Faraday Trans 81:2131, 1985.
52. WR Fawcett, G Liu, TE Kessler. J Phys Chem 97:9293, 1993.

4

Preferential Solvation in Binary Solvent Mixtures

4.1 GENERAL CONSIDERATIONS

Solvation (see Section 1.2) is regarded here in its broadest sense, that is, the molecular interactions that take place in a (liquid) condensed phase that go beyond the geometric proximity of the molecules of the two components dictated by the density. These interactions may be limited to van der Waals forces or may include dipole–dipole or dipole–induced dipole (or more generally, multipole) interactions, hydrogen bonding, and donation of electron density to empty molecular orbitals (donor–acceptor interactions).

The molecular considerations of solvation in binary solvent mixtures deal with the near environment around a given molecule of each component. The question is whether the average composition of the surroundings is the same as that of the bulk mixture or is different. In the former case, there is no preferential solvation and the mixture may be expected to behave (nearly) ideally. In the latter, preferential solvation takes place; the surroundings are enriched relative to the nominal composition of the mixture with respect to either the same component or the second one. If it is the same component, then self-interactions of at least one component are more important than the mutual ones and the preferential solvation is generally manifested by a positive enthalpy of mixing. The mixture may then contain oligomers (dimers, trimers, etc.) of one or both of the components, it may show microheterogeneity (see later, if the preference is very pronounced), and in extreme cases the mixture may even be on the verge of splitting into two liquid phases. It will do so if the pressure and/or the temperature can be and is adjusted as required. If the preference is for the other component, generally manifested by a negative enthalpy of mixing, then the mutual interactions overcome the self-interactions of both components and may result in adduct formation of simple stoichiometry (e.g., 1:1, 1:2) but need not do so.

In either case, if stoichiometric species are formed, such as dimers or adducts, the system may be treated in terms of *associated liquid mixtures*; see Section 2.5. In such mixtures, the species still need not interact ideally with each other. Generally, however, definite stoichiometric species are not formed, and other approaches are required to describe the preferential solvation. The interactions that lead to preferential solvation naturally lead to nonideal behavior of the mixtures and to the macroscopic thermodynamic excess functions described in Section 2.3. These functions, in turn, are interpreted in terms of the molecularly based preferential solvation. Spectroscopic and other measures of physical properties of binary solvent mixtures are usually also interpreted in terms of this concept. The molecular approach has, therefore, as its aim to specify quantitatively the composition of the surroundings of molecules of both components. It may also provide information on the interactions that are responsible for the preferential solvation and shed light on the reasons that are the basis for it.

The specification of the composition of the surroundings of molecules of both components in the binary solvent mixture A + B can be done in terms of the *local mole fraction*, x_{AB}^{L}. This quantity describes the mole fraction of component A in the surroundings of a molecule of component B (where both $A \neq B$ and $A = B$ are considered). The extension of the surroundings has to be specified and is not necessarily confined to the first solvation shell around a given molecule. Necessarily, for any pair of kinds of molecules A and B, $x_{AB}^{L} \leq 1$ and

$$x_{AB}^{L} + x_{BB}^{L} = 1 \tag{4.1}$$

Else, the specification of the preferential solvation can be done in terms of the *preferential solvation parameter*, δx_{AB}. This parameter describes the excess or deficiency of molecules of component A in the vicinity of a molecule of component B:

$$\delta x_{AB} = x^L_{AB} - x_A \tag{4.2}$$

This definition has the following corollaries: $\delta x_{AB} = -\delta x_{BB}$ and $\delta x_{AA} = -\delta x_{BA}$ as well as $\delta x_{AB}(x_A = 0) = \delta x_{AB}(x_A = 1) = 0$ and $\delta x_{AB} \leq x_A$ for any pair of kinds of molecules A and B. Because of the experimental uncertainties in the data that lead to the values of δx_{AB}, these values are not significant when $|\delta x_{AB}| \leq 0.01$, so that over the composition range where such small values are encountered negligible preferential solvation takes place.

A different way to describe the preferential solvation is in terms of the *solvent sorting parameter* that arises from the ratio of local mole fractions to the ratio of the nominal mole fractions:

$$p = (x^L_{AB}/x^L_{BB})/(x_A/x_B) = (x^L_{AB}/x_A)/(x^L_{BB}/x_B) \tag{4.3}$$

Values of $p > 1$ denote *preferential mutual solvation*, values of $p < 1$ denote *preferential self-solvation* (of either component A or B or both), whereas $p \approx 1$ (within, say, ± 0.02) denotes absence of preferential solvation.

Two important questions remain to be discussed before the information conveyed by the various modes of description of the preferential solvation can be assessed. One is the spatial extent of the preferential solvation from a given molecule in the binary mixture. The question is whether the preferential solvation pertains only to the first solvation shell or can also be "felt" and quantitatively measured in further solvation shells and, if so, how far. Most spectroscopic and diffraction (radiation scattering) methods tend to confine their information to the first solvation shell, i.e., the immediate surroundings of the molecule under discussion. Other methods may pertain to a wider environment but not necessarily specify how wide. If the self-preferences of the components extend over sufficiently wide regions in the liquid, *microheterogeneity* results, although the mixture remains macroscopically a single liquid phase. An example of this behavior is the water + acetonitrile system at room temperature. Only if the molecular interactions push the self-preferences further than microheterogeneity may partial immiscibility—phase splitting and liquid–liquid equilibria—be manifested. This process is illustrated by the water + acetonitrile system when cooled below 272.0 K. This tendency culminates in nearly total immiscibility, where the components are hardly soluble at all in one another, as is the case for, say, water and hexane.

The other question that ought to be considered in the present connection is whether the occurrence of preferential solvation in a binary solvent mixture can be predicted from measurable properties of the neat components. That is, can suitable parameters (that may be temperature and pressure dependent) be ascribed to any solvent that accompany it to binary mixtures with any other solvent, from which the nature and extent of preferential solvation can be deduced without any measurements on the mixture being necessary? If the answer to this question were affirmative, then the excess thermodynamic functions should also be predictable. This would be of great pragmatic importance in chemical engineering practice. Unfortunately, our knowledge so far is still a good distance away from this goal.

4.2 DESCRIPTIVE EXPRESSIONS

Descriptive expressions of the thermodynamic excess functions (Section 2.3) and other properties of the mixtures (Section 2.2) lack the ability to predict these functions from the properties of the neat solvents; they only describe them analytically well within the experimental errors of the data. The inability of prediction generally also extends to the preferential solvation. Nevertheless, the application of these expressions to vapor–liquid equilibria (VLE), that is, partial and total vapor pressures and vapor compositions, and to liquid–liquid equilibria (LLE), that is, liquid immiscibility regions, as functions of the nominal mole fraction composition of a binary liquid mixture is of great practical importance. Nonideality of the vapor phase, in the case of VLE, has to be taken into account by means of virial coefficients of the vapor mixtures when the vapor pressures are appreciable. Thus, description of the excess Gibbs energy (and enthalpy, if nonisothermal conditions are to be considered) of binary solvent mixtures by means of the Redlich–Kister and Margules expressions (Section 2.3), although adequate for the accurate calculation of activity coefficients, vapor pressures, etc., does not provide information on the preferential solvation.

It is tempting to use Wilson's expression [1] for the molar excess Gibbs energy for this predictive purpose:

$$G^E/RT = -x_A \ln(x_A - \Lambda_{AB}x_B) - x_B \ln(x_B - \Lambda_{BA}x_A) \qquad (4.4)$$

The expression has two positive unequal parameters, Λ. The parameters Λ_{AB} and Λ_{BA} are related to the interaction energies, e, between the molecules of A and B and to their (partial) molar volumes, V, by

$$\Lambda_{BA} = 1 - (V_B/V_A)\, \exp[-(e_{AB} - e_{AA})/k_B T] \qquad (4.5a)$$

$$\Lambda_{AB} = 1 - (V_A/V_B)\, \exp[-(e_{BA} - e_{BB})/k_B T] \qquad (4.5b)$$

where $e_{AB} = e_{BA}$. The local mole fractions x_{AB}^L should then be related to these interaction energies and the solvent sorting parameter p should be obtained as

$$p = (x_{AB}^L/x_{BB}^L)/(x_A/x_B) = \exp(-e_{AB}/k_B T)/\exp(-e_{BB}/k_B T) \qquad (4.6)$$

However, although only two parameters, Λ_{BA} and Λ_{AB}, are needed to describe the VLE data and are derivable from them, there are three independent interaction energies, $e_{AB} = e_{BA}$, e_{AA}, and e_{BB}, to which these two parameters are related via Eq. (4.5). The two interaction energies that are required for the calculation of the solvent sorting parameter p from Eq. (4.6) cannot be disentangled from Eq. (4.5). Hence, p cannot be derived from the experimental VLE data by this route.

Some other expressions, however, do provide the desired information. Renon and Prausnitz [2] derived the nonrandom two-liquid (NRTL) expressions for both LLE and VLE. Their expression for the molar excess Gibbs energy is

$$\begin{aligned} G^E/RT = x_A x_B \{ & [\tau_{BA}\, \exp(-\alpha\tau_{BA})]/[x_A + x_B\, \exp(-\alpha\tau_{BA})] \\ & + [\tau_{AB}\, \exp(-\alpha\tau_{AB})]/[x_B + x_A\, \exp(-\alpha\tau_{AB})] \} \end{aligned} \qquad (4.7)$$

with three parameters, α, τ_{AB}, and τ_{BA}. The values of the nonrandomness parameter α are empirical, generally between 0.2 and 0.6, and the τ parameters are related to differences in the interaction Gibbs energies g:

$$\tau_{AB} = (g_{AB} - g_{BB})/RT \qquad (4.8a)$$

$$\tau_{BA} = (g_{BA} - g_{AA})/RT \qquad (4.8b)$$

where, as in Wilson's approach, $g_{AB} = g_{BA}$. The local mole fractions are then given by

$$x_{BA}^L = x_B\, \exp(-\alpha\tau_{BA})/[x_A + x_B\, \exp(-\alpha\tau_{BA})] \qquad (4.9a)$$

$$x_{AB}^L = x_A\, \exp(-\alpha\tau_{AB})/[x_B + x_A\, \exp(-\alpha\tau_{AB})] \qquad (4.9b)$$

The expression for the excess Gibbs energy of mixing can then be rewritten in terms of the local mole fractions as:

$$G^E/RT = x_A \tau_{BA} x_{BA}^L + x_B \tau_{AB} x_{AB}^L \tag{4.10}$$

Finally, the solvent sorting parameter p can be calculated from Eqs. (4.9) and (4.3), remembering that $x_{AB}^L + x_{BB}^L = 1$ and $x_{BA}^L + x_{AA}^L = 1$.

Marcus [3] advocated the quasi-lattice quasi-chemical (QLQC) model for binary solvent mixtures, which is an extension of Guggenheim's model [4]. It is closely related to the QLQC approach to the preferential solvation of solutes in binary solvent mixtures discussed in Chapter 5. It bears some relation to the UNIQUAC (*universal quasi-chemical*) [5] model for the excess Gibbs energies of VLE and LLE but permits the evaluation of the local mole fractions of the components that the UNIQUAC approach does not. According to the QLQC model, the binary solvent mixture is characterized by a quasi-lattice with a lattice parameter Z, in which all the lattice sites are occupied by molecules of either component. The lattice parameter Z specifies the number of neighbors each molecule has. In an extension of the method, the lattice parameter is permitted to vary with the composition of the mixture, to take partly into account discrepancies in the sizes of the molecules. Only pairwise interactions of the molecules are considered (so that this model pertains only to the first solvation shell around each molecule). The interaction energies e_{AA}, e_{BB}, and $e_{AB} = e_{BA}$ and the internal degrees of freedom of each molecule are taken to be independent of the natures of the neighbors a molecule may have and of their numbers. Thus, a cooperative effect of two neighboring molecules of a given molecule is excluded by the model. No volume change on mixing is allowed for, so that the Gibbs and Helmholtz energies are set equal to each other.

The quasi-chemical assumption of the QLQC model is derived from the grand partition function that obeys the rules set down for the quasi-lattice [4] and specifies the numbers N and kinds of nearest neighbor pairs of molecules in the mixture:

$$N_{AB}^2/N_{AA}N_{BB} = 4\exp[(e_{AA} + e_{BB} - 2e_{AB})/k_B T] \tag{4.11}$$

This formulation has the form of the equilibrium constant for the exchange of nearest neighbors, $AA + BB \rightleftharpoons 2AB$, hence the appellation quasi-chemical. The total number of nearest neighbors in the mixture is

$$L = (Z/2)(N_A + N_B) = N_{AB} + N_{AA} + N_{BB} \tag{4.12}$$

Hence, $ZN_A = N_{AB} + 2N_{AA}$, $N_{AA}/L = x_A - N_{AB}/2L$, and similarly for ZN_B and N_{BB}/L. Writing for short $\Delta e_{AB} = e_{AA} + e_{BB} - 2e_{AB}$, it follows from these relationships and Eq. (4.11) that

$$N_{AB}/2L = [1-\{1 - 4x_A x_B(1 - \exp(-\Delta e_{AB}/k_B T)\}^{1/2}]$$
$$/[2(1 - \exp(-\Delta e_{AB}/k_B T)] \tag{4.13}$$

The excess chemical potentials and molar excess Gibbs energy of mixing are obtained from [6]

$$\mu_A^E = (ZRT/2) \ln(N_{AA}/Lx_A^2) \tag{4.14a}$$

$$\mu_B^E = (ZRT/2) \ln(N_{BB}/Lx_B^2) \tag{4.14b}$$

and

$$G^E/RT = (Z/2)[x_A \ln(N_{AA}/Lx_A^2) + x_B \ln(N_{BB}/Lx_B^2)] \tag{4.15}$$

On consideration of the excess molar Gibbs energy of the equimolar mixture, $G_{(x=0.5)}^E$, algebraic manipulation of Eqs. (4.11) and (4.15) leads [3] to an explicit expression for Δe_{AB} as a function of $G_{(x=0.5)}^E$ and Z:

$$\exp(\Delta e_{AB}/k_B T) = \{2 \exp[-2G_{(x=0.5)}^E/ZRT] - 1\}^2 \tag{4.16}$$

Further algebraic manipulation yields an expression for the entire excess Gibbs energy of mixing curve:

$$G^E = (ZRT/2)[x_A \ln\{(x_A - (1 - Q)/2P)/x_A^2)\}$$
$$+ x_B \ln\{(x_B - (1 - Q)/2P)/x_B^2)\}] \tag{4.17}$$

where $P=1 - \exp(- \Delta e_{AB}/k_B T)$ and $Q=(1 - 4x_A x_B P)^{1/2}$. This curve is, thus, determined by the single experimental value of $G_{(x=0.5)}^E$ [from Eq. (4.16)] and the lattice parameter Z as a fitting parameter.

The local mole fractions of, say, molecules of A around a molecule of B can now be specified as

$$x_{A(B)}^L = N_{AB}/(2N_{BB} + N_{AB}) = (N_{AB}/2L)/x_B \tag{4.18}$$

where $N_{AB}/2L$ is obtained from Eq. (4.12), and the parentheses around the subscript B serve to distinguish this version of the local mole fraction, which pertains to the first solvation shell only according to the QLQC premises, from other versions.

For the QLQC model, thus, the experimental value of the molar excess Gibbs energy of mixing of the equimolar mixture, $G_{(x=0.5)}^E$, and the lattice parameter Z as a fitting parameter of the $G^E(x)$ curve uniquely determine the local environments of the molecules of the two components and the preferential solvation in the system. In its simplest form it can be used only for rather symmetrical $G^E(x)$ curves.

The model has, however, been extended [3] to include a composition-dependent lattice parameter, $Z(x)$. This arises if there is a large disparity of the sizes of the molecules of the two components. Therefore, in some manner this extension plays the role of the configurational part of the UNIQUAC expression for the excess Gibbs energy, Eq. (2.27). Different values of the lattice parameter, Z_A and Z_B, are assigned to the pure components and a weighting parameter w is used for the mixture:

$$Z(x) = wZ_A + (1 - w)Z_B \tag{4.19}$$

Because only pairwise interactions of nearest neighbors are taken into account, the configurational energy of the pure component A, set equal to the vaporization energy $\Delta_v H_A - RT$, should be given by $(Z_A N_{Av}/2)e_{AA}$ and similarly for pure component B. The ratio $r = Z_A/Z_B$ is then used to define an energy quantity Δ:

$$\begin{aligned}\Delta &= [e_{AA}/Z_A - e_{BB}/Z_B]/k_B T = (2/RTZ_A^2)[\Delta_v H_A - r^2 \Delta_v H_B \\ &\quad + (r^2 - 1)RT]\end{aligned} \tag{4.20}$$

This quantity, in turn, determines the weighting parameter w:

$$w = x_B^{1/r} \exp(\Delta)[x_A^{1/r} + x_B^{1/r} \exp(\Delta)] \tag{4.21}$$

The lattice parameter at the equimolar composition is

$$Z_{(x=0.5)} = Z_A[1 + (1/r) \exp(\Delta)]/[1 + \exp(\Delta)] \tag{4.22}$$

and should be used in Eq. (4.16) for obtaining the exchange energy Δe_{AB}. Thus, with the two fitting parameters Z_A and r and the experimental value of $G^E_{(x=0.5)}$ the entire $G^E(x)$ curve, even when (moderately) skew, can be fitted and the local mole fractions x^L can be evaluated [3]; see correction in Ref. 7

The QLQC method should, in principle, be applicable to the calculation of the molar excess enthalpy of mixing from $H^E = G^E - TdG^E/dT$, using Eq. (4.17) and the required temperature derivatives of the variables there. If the lattice parameter Z is taken as temperature and composition independent, then only the derivatives of the quantities P and Q are required, and the resulting expression is

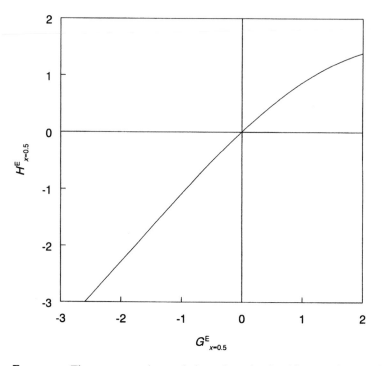

FIGURE 4.1 The excess molar enthalpy of equimolar binary solvent mixtures, $H^E_{(x=0.5)}$, as a function of the excess molar Gibbs energy of such mixtures, calculated according to the QLQC method for 298.15 K and $Z = 4$.

$$H^E = -(ZRT/2)[(\Delta e_{AB}/k_B T) \exp(-\Delta e_{AB}/k_B T)/(1 - \exp(-\Delta e_{AB}/k_B T))]$$
$$\times \{ x_A(1 - Q - 2x_A x_B P/Q)(1 - Q - 2x_A P)^{-1}$$
$$+ x_B(1 - Q - 2x_A x_B P/Q)(1 - Q - 2x_B P)^{-1} \} \qquad (4.23)$$

This expression pertains only to rather symmetrical G^E and H^E curves. The dependence of the calculated $H^E_{(x=0.5)}$ on $G^E_{(x=0.5)}$ for $Z = 4$ is shown in Fig. 4.1. However, in the more general case, Z depends on the composition, and if the preceding considerations regarding Z are accepted, these also involve the temperature dependence of Z through Eqs. (4.20) to (4.22). It turns out, however, that the values of $H^E_{(x=0.5)}$ calculated according to such considerations do not agree well with the experimental values except in a few cases; hence this derivation is not presented.

A quite different approach to preferential solvation in binary solvent mixtures is based on the Kirkwood and Buff theory [8] of the relationship of the space integral of the pair correlation function, $g_{ij}(r)$ to thermodynamic data. The function $g_{ij}(r)$ expresses the probability that given (the center of) a molecule of component i at a certain position in the mixture, (the center of) a molecule of species j is in an infinitesimal volume element at a distance r. Both $j = i$ and $j \neq i$ can be considered in this connection. The (Kirkwood–Buff) integral is

$$G_{ij} = \int_0^\infty [g_{ij}(r) - 1] 4\pi r^2 \, dr \qquad (4.24)$$

In liquids in general and solvent mixtures in particular, the function $g_{ij}(r)$ differs from unity only in a region a few molecular diameters away from the center of molecule i, because no long-range order is maintained. Therefore, the main contribution to G_{ij} comes from this limited region, called the *correlation region*. Beyond the maximal extension of the correlation region, at $r > R_{cor}$, the molecule i at the origin no longer has any effect on the presence or absence of any particular molecule in the volume element, so that the probability $g_{ij}(r) = 1$. The volume of the correlation region is $V_{cor} = (4\pi/3)R_{cor}^3$. The product of the (bulk) number density of component j, ρ_j, and G_{ij} is the excess (or deficiency) of molecules of j in the correlation volume around a molecule of i. In binary solvent mixtures A + B, there are three Kirkwood–Buff integrals that should be considered: G_{AA}, G_{BB}, and G_{AB}. For systems mixing ideally, $G_{AA} + G_{BB} - 2G_{AB} = 0$, but in nonideal mixtures the three integrals are not simply related to one another. Knowledge of their dependence on the composition provides information on the preferential solvation in the mixture.

It was shown by Ben-Naim [9] that the Kirkwood–Buff integrals in binary solvent mixtures A + B can be calculated from thermodynamic data as follows:

$$G_{AB} = RT\kappa_T - V_A V_B / VQ \qquad (4.25a)$$

$$G_{AA} = RT\kappa_T - (x_B/x_A)V_B^2 / VQ - V/x_A \qquad (4.25b)$$

$$G_{BB} = RT\kappa_T = (x_A/x_B)V_A^2 / VQ - V/x_B \qquad (4.25c)$$

using the modified form suggested by Matteoli and Lepori [10], where

$$Q = x_j[\partial(\mu_i/RT)/\partial x_i]_{T,P} = x_j[\partial \ln a_i/\partial x_i]_{T,P} = 1 + x_j[\partial \ln f_i/\partial x_j]_{T,P}$$

$$= 1 + x_A x_B[\partial^2(G^E/RT)/\partial x_A^2]_{T,P} \qquad (4.26)$$

The evaluation of the function Q, Eq. (4.26), requirs the first derivative of the activity coefficient f or of the activity a of one of the components or the second derivative of the excess Gibbs energy of mixing, G^E, with respect to the composition. The calculation of the Kirkwood–Buff integrals, Eqs. (IV. 25), also requires the isothermal compressibility of the mixture, κ_T; the partial molar volumes of the components, V_A and V_B; and the total molar volume, $V = x_A V_A + x_B V_B$. The Kirkwood–Buff integrals determined for aqueous methanol, ethanol, and n-propanol are shown in Fig. 4.2 on the same scale for comparison. The large positive G_{AA} and negative G_{AB} values for A =water and B = n-propanol should be noted. These are commented on further in the following.

It should be noted that Kirkwood–Buff integrals can be obtained not only from thermodynamic data but also from structural data, namely from molecular number density fluctuations studied by light scattering or other methods. According to Kato [11], the derivative function Q defined in Eq. (4.26) and appearing in the thermodynamic expressions for the Kirkwood–Buff integrals, Eqs. (4.25), is related to the mean square concentration fluctuation, $\langle(\Delta x)^2\rangle$, as follows:

$$Q = x_A x_B/N_{Av}\langle(\Delta x)^2\rangle \qquad (4.27)$$

The quantity $\langle(\Delta x)^2\rangle$ in turn is obtained from the Rayleigh scattering data. The procedure can be reversed in order to obtain the mean square concentration fluctuation from the Kirkwood–Buff integrals:

$$N_{Av}\langle(\Delta x)^2\rangle = x_A x_B[1 + x_A x_B(G_{AA} + G_{BB} - 2G_{AB})/V] \qquad (4.28)$$

If self-preference of one of the components or both takes place, then $N_{Av}\langle(\Delta x)^2\rangle$ is positive, whereas if mutual preference predominates this quantity is negative. See also Section 3.1, for fluctuations derived from diffraction data.

The local mole fraction of molecules of component A around a molecule of the same component is given according to Ben-Naim [12] by

$$x_{AA}^L = (\rho_A G_{AA} + V_{corr})/(\rho_A G_{AA} + \rho_B G_{BA} + 2V_{corr}) \qquad (4.29a)$$

Similarly, the local mole fraction of component A around a molecule of component B is

$$x_{BA}^L = (\rho_A G_{BA} + V_{corr})/(\rho_A G_{BA} + \rho_B G_{BB} + 2V_{corr}) \qquad (4.29b)$$

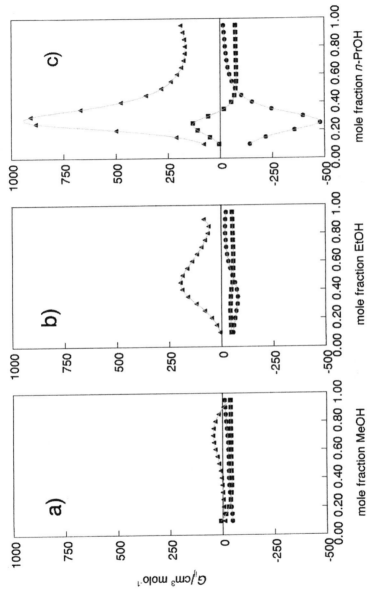

Figure 4.2 The Kirkwood–Buff integrals G_{ij} of aqueous (component A) methanol (a), ethanol (b), and n-propanol (c) (components B): G_{AB}(●), G_{AA}(▲), G_{BB}(■).

Therefore, the three integrals G_{AA}, G_{BB}, and G_{AB} as well as the quantity V_{corr} are required in order to calculate the two independent quantities $x^L_{AA} = 1 - x^L_{AB}$ and $x^L_{BA} = 1 - x^L_{BB}$.

4.3   THE PREFERENTIAL SOLVATION PARAMETERS

The local mole fractions calculated from either the QLQC or the inverse Kirkwood–Buff integral (IKBI) method can be transformed into *preferential solvation parameters* by subtraction of the bulk mole fraction of the component surrounding the molecule of the component whose preferential solvation is discussed. In the case of the QLQC method, the derived quantity is, e.g., $\delta x_{A(B)} = x^L_{A(B)} - x_A$, where $x^L_{A(B)}$ is obtained from Eq. (4.18). In the binary mixtures, necessarily, $\delta x_{B(B)} = -\delta x_{A(B)}$, $\delta x_{B(A)} = (x_B/x_A)\delta x_{A(B)}$, and $\delta x_{A(A)} = -\delta x_{B(A)}$.

The Kirkwood–Buff integrals, calculated from the thermodynamic data, can also be used for the calculation of the local mole fractions and the preferential solvation parameters in the binary solvent mixture. The latter quantities are [12]

$$\delta x_{AA} = x^L_{AA} - x_A = x_A x_B (G_{AA} - G_{BA})/[x_A G_{AA} + x_B G_{AB} + V_{corr}]$$

(4.30a)

$$\delta x_{AB} = x^L_{AB} - x_A = x_A x_B (G_{AB} - G_{BB})/[x_A G_{AB} + x_B G_{BB} + V_{cor}]$$

(4.30b)

The two quantities δx_{AA} and δx_{AB} are independent of each other, but $\delta x_{BA} = -\delta x_{AA}$ and $\delta x_{BB} = -\delta x_{AB}$. It is still necessary to specify a value for the correlation volume V_{cor} or the correlation distance R_{cor}. An iterative calculation of these quantities is required because they depend, in turn, on the local composition of the environment of the species A (or B). The mean diameter of a solvent molecule i (A or B) was suggested by Kim [13] to be given by

$$\sigma_i/nm = -0.085 + 0.1363(V^\circ_i/cm^3 \, mol^{-1})^{1/3}$$

(4.31)

where V°_i is the molar volume of pure component i. The correlation distance up to the nth solvation shell around a molecule of component A was therefore given by Marcus [14] as

$$R_{cor(A)}/nm = -0.085m + (0.1363/2)(V^\circ_A)^{1/3}$$
$$+ 0.1363(m - 0.5)[x^L_{AA}V^\circ_A + (1 - x^L_{AA})V^\circ_B]^{1/3} \quad (4.32a)$$

and around a molecule of component B as

$$R_{cor(B)}/nm = -0.085m + (0.1363/2)(V_B^\circ)^{1/3}$$
$$+ 0.1363(m - 0.5)[x_{AB}^L V_A^\circ + (1 - x_{AB}^L)V_B^\circ]^{1/3} \quad (4.32b)$$

The parameter m takes the values 1.5, 2.25, and 2.9 for the first three solvation shells and $0.7(n+1)$ for the nth shell from the fourth onward, to account for the mean geometry of the packing of solvent molecules in the spherical shells and for the partial penetration of a solvent molecule from the nth shell into the $(n-1)$th shell [14]. Hence, the preferential solvation parameters obtained from the IKBI method can describe the preferential solvation parameters in farther regions of space around a molecule in the binary solvent mixture than those obtained by the QLQC method. This should be borne in mind when results from the two methods are compared.

A feature to be noted in the application of Eq. (4.30) is that even for ideal mixtures (where $G^E = 0$ and $V^E = 0$) the δx_{ij} are nonzero, as if preferential solvation existed in them. This arises from the differences in the molar volumes of the pure components, resulting in Kirkwood–Buff integral curves G_{AA}, G_{BB}, and G_{AB} for ideal mixtures that are parallel, but the former two are shifted by $|V_B^\circ - V_A^\circ|$ from the G_{AB} curve and satisfy $G_{AA} + G_{BB} - 2G_{AB} = 0$. To overcome this, Matteoli [15] suggested the use of "volume-corrected" values, $\Delta G_{AB} = G_{AB} - G_{AB}^{id}$ (and similarly for ΔG_{AA} and ΔG_{BB}), where the "ideal" G_{AB}^{id} are calculated from Eq. (4.25) by setting $Q = 1$ and $V^E = 0$. Then volume-corrected preferential solvation parameters $\delta x'$ are calculated from Eq. (4.30) using ΔG instead of G. The actual values of the local mole fractions and the preferential solvation parameters still depend on G rather than on ΔG because of the differences in the volumes of the components, but the interpretation in terms of intermolecular interactions should be done with regard to $\delta x'$. Values of δx_{AB} for ideal systems with $V_A^\circ = 20$ cm^{3}mol^{-1} (near that of water) and various molar volumes of the cosolvent are shown in Fig. 4.3. Of course, $\delta x_{AB}' = 0$ for these ideal systems.

In view of the extensive use of the Kirkwood–Buff integral method for the evaluation of the preferential solvation in binary solvent mixtures, it is instructive to examine the accuracy of the values of G_{ij} that can be achieved. It is assumed that the tests requiring that $\lim G_{AA}(x_B \to 0) = RT\kappa_{TA} - V_A^\circ$ and $\lim G_{BB}(x_A \to 0) = RT\kappa_{TB} - V_B^\circ$ have shown that the calculation procedure has been carried out satisfactorily. Generally, $RT\kappa_T$ is of the order of 1 cm^3 mol^{-1} and is small with respect to V, and $|G_{ij}|$. The latter quantity is of the order of tens or hundreds of cm^{3}mol^{-3} in

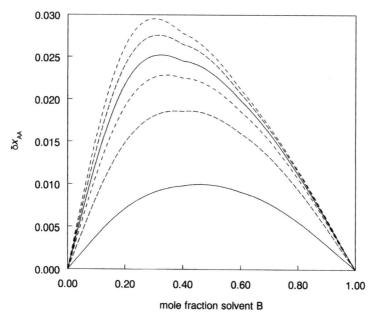

FIGURE 4.3 The preferential solvation parameters δx_{AA} of solvent A ($V_A = 20\,\text{cm}^3\,\text{mol}^{-1}$) in its *ideal* mixtures with solvents B of increasing molar volumes (from the bottom): 30, 50, 70, 90, 120, and $180\,\text{cm}^3\,\text{mol}^{-1}$. From Ref. [21] with permission.

homogeneous binary mixtures, although it may change sign and be near zero in a small composition region. Inaccuracies in $\kappa_T(x)$ are, therefore, of minor consequence. Hence, the adiabatic compressibilities, $\kappa_S(x)$, can be used instead of the isothermal ones, or $\kappa_T(x)$ may even be linearly interpolated between $\kappa_{T(x=0)}$ and $\kappa_{T(x=1)}$ without serious loss of accuracy. The most troublesome quantity is Q, which depends on derivative functions. The excess Gibbs energy of mixing G^E is generally obtained from the activities or activity coefficients in the first place, and for the latter quantities only the first derivative has to be calculated, in contrast to the second derivative of G^E, hence it is obvious what data should be preferred. Still, if in the Redlich–Kister representation of G^E the leading term $g_0 \approx RT$, the value of G_{AB} of the equimolar mixture is $\approx -100\,\text{cm}^3\,\text{mol}^{-1}$ with a probable error of $2\,\text{cm}^3\,\text{mol}^{-1}$. However, if the leading term is as large as $g_0 = 1.9RT$, signifying a mixture on the verge of splitting into two immiscible liquid phases, then for the equimolar mixture $G_{AB} \approx -1000\,\text{cm}^3\,\text{mol}^{-1}$ with a probable error of $380\,\text{cm}^3\,\text{mol}^{-1}$, as shown by Zaitsev et al. [16,17]. Very large positive values (of the order of 10^4 cm^3 mol^{-1}) then result for G_{AA} and/or G_{BB}, depending on which of the components self-associates preferentially; see Fig. 4.2.

The Kirkwood–Buff integrals have been calculated for a large number of binary aqueous mixtures of organic solvents ([10,14,18–20] among others) as well as nonaqueous ones. The values of the volume-corrected preferential solvation parameters $\delta x_{AA}'$ and $\delta x_{AB}'$ for the cosolvents of water on the present list and a few additional ones, forming completely miscible binary aqueous mixtures, are shown in Table 4.1 for the temperature 298.15 K, unless otherwise noted. The preferential solvation parameter values for mutual interactions, $\delta x_{AB}'$, are generally lower (in the absolute sense) than those of the self-interaction parameters of the water, $\delta x_{AA}'$. Cosolvents with sufficiently large hydrophobic groups have positive values of $\delta x_{AA}'$ and negative ones of $\delta x_{AB}'$. The larger these hydrophobic groups, the larger is $\delta x_{AA}'$. Small, hydrophilic co-solvents with strong donor functional groups have positive mutual preferential solvation parameters, $\delta x_{AB}'$, with water and negative water self-interaction ones, $\delta x_{AA}'$. Values of $|x_{AA}'|$ and $|\delta x_{AB}'| < 0.01$ are probably insignificant.

Following are detailed descriptions [21] of the interactions noted in the systems included in Table 4.1, where water is designated by subscript W and the cosolvent by subscript S.

Aqueous hydrogen peroxide. This system is close to ideal: for the disfavored water self-interactions the minimal $\delta x_{WW}' = -0.013$ at $x_S = 0.70$, whereas the favored mutual interaction has a maximal $\delta x_{WS}' = 0.009$ near $x_S = 0.35$, both extrema being quite small, and preferential solvation is hardly significant.

Aqueous methanol. For the water self-interactions the maximal $\delta x_{WW}' = 0.037$ at $x_S = 0.65$, whereas for the mutual interaction the minimal $\delta x_{WS}' = -0.017$ near $x_S = 0.10$. The maximal $\delta x_{WW}'$ value can be compared with those shown next for aqueous ethanol and *n*-propanol: 0.16 and 0.42, showing the expected gradual change.

Aqueous ethanol. The volume-corrected preferential solvation parameter curves show that the water tends to aggregate near a water molecule, with a maximum of $\delta x_{WW}' = 0.16$ near $x_S = 0.45$, and that it tends to avoid slightly the vicinity of the ethanol, with a minimum of $\delta x_{WS}' = -0.04$ near $x_S = 0.35$.

Aqueous n-propanol. The volume-corrected preferential solvation parameter curves show that the water tends much more than in the case of aqueous ethanol to aggregate near a water molecule,

TABLE 4.1 First Solvation Shell Volume-Corrected Preferential Solvation Parameters $10^3\ \delta x'_{AA}$ (First Row) and $10^3\ \delta x_{AB}'$ (Second Row) for Aqueous Mixtures (A = Water and B = Cosolvent) at 298.15 K Obtained from the Kirkwood–Buff Integrals, as a Function of x_B

Cosolvent	$x_B \rightarrow$	0.15	0.35	0.50	0.65	0.85	Ref.
Hydrogen peroxide		−2	−7	−11	−13	−11	a
		6	9	7	5	2	
Methanol		15	16	26	37	5	c
		−15	−3	−3	−2	1	
Ethanol		23	141	139	60	11	e
		−18	−36	−18	−4	−1	
1-Propanol		151	405	182	81	31	g
		−100	−117	−20	−5	0	
2-Propanol		28	131	167	126	45	a
		−21	−31	−21	−8	−1	
2-Methyl-2-propanol		140	400	195	67	13	e, i
		−59	−100	−15	−4	−1	
2,2,2-Trifluoroethanol		180	235	110	48	−1	f
		−142	−50	−12	−2	0	
1,1,1,3,3,3-Hexafluoro-2-propanol		230	43	−22	−40	−33	f
		−40	−3	3	2	1	
1,2-Ethanediol		−2	−10	−12	−9	1	a, h
		4	5	4	2	0	
Glycerol		5	10	10	12	12	b, j
		−5	−2	−1	−1	0	
2-Aminoethanol		0	−18	−36	−46	−33	g
		2	6	7	5	2	
N-Methyl-2-aminoethanol		−4	−12	−29	−41	−35	e
		3	3	4	3	1	
N,N-Dimethyl-2-aminoethanol		8	0	−12	−18	−13	e
		−2	1	2	2	1	
N,N-Diethanolamine		−14	−29	−33	−32	−20	h
		6	5	3	2	1	
N,N,N-Triethanolamine (273 K)		−25	−45	−50	−48	−32	h
		6	5	3	2	0	
2-Methoxyethanol (343 K)		6	40	69	55	8	e
		3	−2	−5	−2	0	
2-Ethoxyethanol		140	84	14	−4	10	e
		−57	−5	1	1	0	
2-Butoxyethanol		69	44	24	18	4	e
		−19	−4	−1	0	0	

TABLE 4.1 Continued

Cosolvent	$x_B \rightarrow$	0.15	0.35	0.50	0.65	0.85	Ref.
Tetrahydrofuran		162	431	333	268	152	e
		− 102	− 151	− 33	− 14	− 3	
Dioxane		35	196	448	568	160	e
		− 12	− 32	− 50	− 35	− 3	
Acetone		51	240	313	269	105	e
		− 23	− 45	− 31	− 13	− 1	
Formic acid		− 15	− 20	− 12	− 19	− 7	a
		33	15	6	5	1	
Acetic acid		11	42	50	42	20	a
		− 15	− 16	− 9	− 4	0	
Triethylamine (283 K)		− 23	− 36	− 40	− 38	− 27	a
		4	3	2	2	1	
Pyridine		129	60	27	28	54	e
		− 79	− 9	− 1	0	0	
Piperidine		98	− 2	− 10	21	29	g
		− 33	3	3	1	0	
Acetonitrile		119	559	481	335	134	d
		− 116	− 326	− 108	− 37	− 5	
Formamide		9	7	5	4	2	a
		− 13	− 2	− 1	0	0	
N-Methylformamide		− 10	− 38	− 44	− 51	− 49	a
		8	12	8	5	2	
N-Methylacetamide (333 K)		− 2	0	2	2	1	a
		− 1	− 1	0	1	0	
N,N-Dimethylformamide		− 6	− 16	− 15	− 6	12	a
		5	5	4	2	0	
N-Methylpyrrolidin-2-one		3	− 10	− 10	− 3	− 6	e
		0	2	2	1	1	
Hexamethyl phosphoric triamide		11	38	49	46	24	a
		− 3	− 2	− 2	− 1	0	
Dimethylsulfoxide		− 4	− 25	− 43	− 55	− 55	e
		0	6	6	5	1	
Tetramethylenesulfone (303 K)		64	308	304	174	53	a
		− 32	− 57	− 31	− 10	− 1	

The references pertain to the published Kirkwood–Buff integrals.
[a]Y Marcus. J Chem Soc Faraday Trans 86:2215, 1990.
[b]Y Marcus. Phys Chem Chem Phys 2:4891, 2000.
[c]Y Marcus. Phys Chem Chem Phys 1:2975, 1999.
[d]Y Marcus, Y Migron [7].
[e]Y Marcus [21].
[f]MJ Blandamer et al. [18–20].
[g]E Matteoli, L Lepori [10].
[h]Y Marcus. J Chem Soc Faraday Trans 91:427, 1995.
[i]Data for 1-butanol also shown in the miscibility region in Ref. g.
[j]Data for tri- and tetraethylene glycol were reported by E Matteoli. J Phys Chem B101:9800, 1997.

with a maximum of $\delta x'_{WW} = 0.42$ in the first solvation shell and $\delta x'_{WW} = 0.22$ in the second, near $x_S = 0.30$. Even in the third solvation shell is the maximal $\delta x'_{WW} > 0.1$, indicating that this system is not far from microheterogeneity. The avoidance of water in the vicinity of a n-propanol molecule is correspondingly pronounced; the minima of $\delta x'_{WS}$ are -0.28 and -0.09 in the first and second solvation shells, near $x_S = 0.25$.

Aqueous i-propanol. The water self-interactions have the maximal $\delta x'_{WW} = 0.165$ at $x_S = 0.50$, whereas for the mutual interaction the minimal $\delta x'_{WS} = -0.031$ near $x_S = 0.30$. The preferential solvation is considerably smaller than for n-propanol and t-butanol (maximal $\delta x'_{WW} = 0.40$) and is comparable to that of ethanol.

Aqueous t-butanol. The volume-corrected preferential solvation parameters for the first and second solvation shells show that water tends as much as in the case of aqueous n-propanol to aggregate near a water molecule, with a maximum of $\delta x'_{WW} = 0.40$ in the first solvation shell and 0.28 in the second, near $x_S = 0.35$ and 0.30, respectively, and in the third solvation shell it is still 0.15. The avoidance of water in the vicinity of a t-butanol molecule is also pronounced; the minima of $\delta x'_{WS}$ are -0.24 and -0.08 in the first and second solvation shells, near $x_S = 0.25$.

Aqueous 1,2-ethanediol (ethylene glycol). This system is, as expected, much more nearly ideal than the monoalcohols. As for aqueous hydrogen peroxide, for the disfavored water self-interactions the minimal $\delta x_{WW}' = -0.012$ at $x_S = 0.50$, whereas the favored mutual interaction has a maximal $\delta x_{WS}' = 0.005$ near $x_S = 0.35$, both extrema being quite small, and preferential solvation is hardly significant.

Aqueous glycerol. As for aqueous ethylene glycol, the system is close to ideal, but the self-interactions of water are now favored over the mutual interactions. The maximal $\delta x_{WW}' = 0.014$ at $x_S = 0.75$, whereas the disfavored mutual interaction has a minimal $\delta x'_{WS} = -0.005$ near $x_S = 0.15$, both extrema being again rather small.

Aqueous 2,2,2-trifluoroethanol. The $\delta x'_{WW}$ and $\delta x'_{WS}$ curves lie in between those of ethanol and n-propanol, with a maximal $\delta x'_{WW} = 0.31$ near $x_S = 0.25$ and a minimal $\delta x'_{WS} = -0.20$ near $x_S = 0.20$.

Aqueous 1,1,1,3,3,3-hexafluoro-i-propanol. The $\delta x'_{WW}$ and $\delta x'_{WS}$ curves are somewhat lower (in absolute value) than those for aqueous 2,2,2-trifluoroethanol.

Aqueous 2-methoxyethanol. The Kirkwood–Buff integrals, $G_{WW}(x_S)$, $G_{WS}(x_S)$, and $G_{SS}(x_S)$, were obtained for 313.15 and 343.15 K but not 298.15 K. It is noted that the temperature has a negligible effect on $G_{WS}(x_S)$ and $G_{SS}(x_S)$ but does affect $G_{WW}(x_S)$, the curve at 343.15 K being broader and somewhat lower than at 313.15 K. The resulting $\delta x_{WS}'$ are, correspondingly, practically identical at the two temperatures, but the $\delta x_{WW}'$ curves differ, that at 343.15 K being again broader and somewhat lower than that at 313.15 K. At the latter temperature the maximal $\delta x_{WW}' = 0.10$ and the minimal $\delta x_{WS}' = -0.008$, both near $x_S = 0.50$. These values are smaller in absolute value than those for ethanol, showing the effect of the additional (etheral) oxygen atom on the interactions.

Aqueous 2-ethoxyethanol. The Kirkwood–Buff integrals, $G_{WW}(x_S)$, $G_{WS}(x_S)$, and $G_{SS}(x_S)$, have the expected behavior, showing more extreme values than the corresponding curves for 2-methoxyethanol in the more water-rich mixtures, the ethyl group being more hydrophobic than the methyl group. The calculated $\delta x_{WW}'$ and $\delta x_{WS}'$ have extrema that are larger than those of 2-methoxyethanol and in more water-rich mixtures: the maximal $\delta x_{WW}' = 0.18$ near $x_S = 0.20$ and the minimal $\delta x_{WS}' = -0.06$ near $x_S = 0.15$.

Aqueous 2-n-butoxyethanol. The derived $\delta x_{WW}'$ and $\delta x_{WS}'$ values are lower than expected for gradual changes from methoxy- through ethoxy- to butoxyethanol but are of the right sign and shape and have the maximal $\delta x_{WW}' = 0.075$ near $x_S = 0.20$ and the minimal $\delta x_{WS}' = -0.019$ near $x_S = 0.15$, i.e., at more water-rich compositions that for the methoxy- and ethoxyethanols, as expected.

Aqueous formic acid. For this very hydrophilic cosolvent the water self-interactions are disfavored, with a minimal $\delta x_{WW}' = -0.022$ at $x_S = 0.30$, whereas the favored mutual interaction has a maximal $\delta x_{WS}' = 0.033$ near $x_S = 0.15$. These values for the extrema are rather small, showing that the system is not far from ideality.

Aqueous acetic acid. This system shows an intermediate magnitude of nonideality, the water–water interactions being favored, with maximal $\delta x_{WW}' = 0.050$ at $x_S = 0.50$, whereas the disfavored mutual water–acid interaction has a minimal $\delta x_{WS}' = -0.018$ at $x_S = 0.25$.

Aqueous propanoic acid. Surprisingly, this system exhibits lower nonideality than the aqueous acetic acid system, the water–water interactions being favored, but with maximal $\delta x'_{WW} = 0.020$ only at $x_S = 0.25$, whereas the disfavored mutual water–acid interaction has a minimal $\delta x'_{WS} = -0.013$ at $x_S = 0.15$.

Aqueous tetrahydrofuran. Considerable water–water association takes place, the maximal $\delta x'_{WW} = 0.33$ near $x_S = 0.30$ for the first solvation shell. This self-association persists to farther shells, Fig. 4.4, and even in the fifth shell it is not completely negligible (maximal $\delta x'_{WW} = 0.028$). The mutual association of water with tetrahydrofuran is correspondingly depressed, with the minimal $\delta x'_{WS} = -0.20$ at $x_S = 0.25$, and this persists again to farther solvation shells.

Aqueous 1,4-dioxane. The $\delta x'_{WW}$ values show even larger self-association of the water than in aqueous tetrahydrofuran, the maximal $\delta x'_{WW} = 0.57$ at $x_S = 0.65$, and again persists to farther solvation shells. However, the depression of the mutual interaction of water and dioxane is not as far reaching as in the case of aqueous tetrahydrofuran, the minimal $\delta x'_{WS} = -0.052$ at $x_S = 0.55$, and hardly persists ($\delta x'_{WS} > -0.007$) in the third solvation shell because of the existence of two oxygen atoms in the dioxane molecule.

Aqueous acetone. The $\delta x'_{WW}$ values are less pronounced than in the ethers, with water–water interactions reaching the maximal $\delta x'_{WW} = 0.31$ near $x_S = 0.5$ and the mutual water–acetone interactions reaching the minimal $\delta x'_{WW} = -0.045$ at $x_S = 0.35$.

Aqueous 2-aminoethanol. The calculated $\delta x'_{WW}$ and $\delta x'_{WS}$ are quite small (the minimal $\delta x'_{WW} = -0.046$ and the maximal $\delta x'_{WS} = 0.007$). The main feature for the system involving the very hydrophilic 2-aminoethanol is that $\delta x'_{WW}$ is negative and $\delta x'_{WS}$ is positive, signifying strong attraction between the water and the cosolvent.

Aqueous N-methyl-2-aminoethanol. The resulting $\delta x'_{WW}$ and $\delta x'_{WS}$ are very similar to those for 2-aminoethanol, but the extrema are even slightly smaller (the minimal $\delta x'_{WW} = -0.043$ and the maximal $\delta x'_{WS} = 0.004$), signifying the expected somewhat lower hydrophilicity of the methyl derivative.

Aqueous N,N-dimethyl-2-aminoethanol. The resulting $\delta x'_{WW}$ and $\delta x'_{WS}$ follow the trend set by the unsubstituted and singly substituted aminoethanols, with $\delta x'_{WW}$ now showing slightly positive values at water-rich compositions, signifying that the self-association of water in this region overcomes the mutual association due to the two methyl groups.

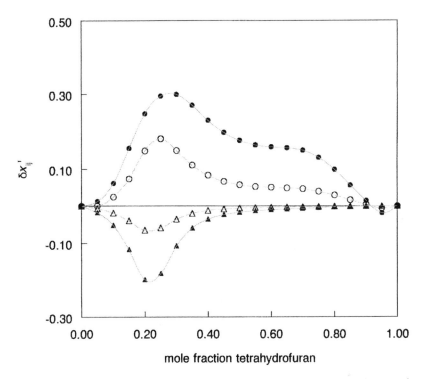

FIGURE 4.4 The volume-corrected preferential solvation parameters $\delta x_{WW'}$ (circles) and $\delta x_{WS'}$ (triangles) in the first (filled symbols) and second (empty symbols) solvation shells in aqueous tetrahydrofuran mixtures at 298.15 K.

Aqueous piperidine. The calculated positive $\delta x'_{WW}$ and negative $\delta x'_{WS}$ are quite small, their absolute values being smaller than 0.1.

Aqueous pyridine. The derived $\delta x'_{WW}$ and $\delta x'_{WS}$ values lead to small ($\delta x' < 0.07$) self- and mutual preferences at $x_S > 0.4$.

Aqueous acetonitrile. This system exhibits large deviations from ideal mixing (immiscibility is noted at ≤ 272.0 K), and preferential solvation that extends over many solvation shells (microheterogeneity) still occurs at 298.15 K. For the water self interactions the maximal $\delta x'_{WW} = 0.56$ at $x_S = 0.35$, whereas for the mutual interaction the minimal $\delta x'_{WS} = -0.37$ near $x_S = 0.30$.

Aqueous formamide. The volume-corrected preferential solvation parameters indicate that this system is nearly ideal. The extrema are very small and preferential solvation is hardly significant. The water self-interactions are favored, but with maximal $\delta x'_{WW} = 0.009$ only at $x_S = 0.15$, and the mutual interactions have a minimal $\delta x'_{WS} = -0.013$ at $x_S = 0.15$.

Aqueous N-methylformamide. The water self-interaction is disfavored in this case, with minimal $\delta x'_{WW} = -0.054$ at $x_S = 0.75$, and the favored mutual interactions have a maximal $\delta x'_{WS} = 0.014$ at $x_S = 0.25$. The better compatibility of this cosolvent than of formamide with water is unexpected, although the extreme values are fairly small in either case.

Aqueous N-methylacetamide. The behavior of this system is practically indistinguishable from ideal behavior, the absolute extreme values of the volume-corrected preferential solvation parameters being < 0.002, so that no preferential solvation takes place.

Aqueous N,N-dimethylformamide. The water self-interactions are disfavored at low cosolvent concentrations, with minimal $\delta x'_{WW} = -0.016$ at $x_S = 0.40$, but somewhat favored at high cosolvent concentrations, reaching $\delta x'_{WW} = -0.012$ at $x_S = 0.90$. The favored mutual interactions have a small maximal $\delta x'_{WS} = 0.006$ at $x_S = 0.30$. Little preferential solvation is noted for this system.

Aqueous N-methylpyrrolidin-2-one. The calculated $\delta x'_{WW}$ and $\delta x'_{WS}$ values are very small and at $x_S > 0.2$ are negative for $\delta x'_{WW}$ and positive for $\delta x'_{WS}$, signifying slight preference for mutual over self-association in the solvation shells, as in the cases of the ethanolamines.

Aqueous hexamethyl phosphoric triamide. The extrema in the volume-corrected preferential solvation parameters are not large, the water self-interactions being favored, but with maximal $\delta x'_{WW} = 0.050$ at $x_S = 0.55$, and the mutual interactions have a minimal $\delta x'_{WS} = -0.003$ at $x_S = 0.20$. Although the phosphoryl group and the three amidic oxygen atoms are exposed to the water, the six methyl groups exert their hydrophobic nature effectively.

Aqueous dimethylsulfoxide. The Kirkwood–Buff integrals lead to rather small negative $\delta x'_{WW}$ and positive $\delta x'_{WS}$ values. As expected, the self-interaction of the water in the solvation shells is depressed, the minimal $\delta x'_{WW} = -0.058$ at $x_S = 0.75$, whereas the mutual interaction, with a maximal $\delta x'_{WS} = 0.006$ near $x_S = 0.45$, is quite small.

Aqueous tetramethylenesulfone (sulfolane). The volume-corrected preferential solvation parameters (for 303.15 K) show that the water self-interactions are strongly favored, with maximal $\delta x'_{WW} = 0.34$ at $x_S = 0.40$, and the disfavored mutual interactions have a minimal $\delta x'_{WS} = -0.06$ at $x_S = 0.30$. The pronounced preferential solvation extends over several solvation shells and is similar in magnitude to that noted in aqueous acetone or tetrahydrofuran.

Of the systems considered here, only the aqueous *n*-propanol, *t*-butanol, and tetramethylenesulfone mixtures approach aqueous acetonitrile in exhibiting that large preferential solvation. In these systems appreciable preferential solvation prevails over several solvation shells around a given molecule, microheterogeneity being a consequence of this fact. The temperatures at which the latter two systems are presented here are, however, only slightly higher than the melting points of the neat cosolvents, a fact that ought to affect the behavior of these two systems.

A somewhat less comprehensive study of these aqueous (A) cosolvent (B) mixtures was made by means of the QLQC method. The excess Gibbs energy of mixing curves, $G^E(x_B)$ could be fitted well by means of Eq. (4.17) using a variable lattice parameter $Z(x_B)$, Eq. (4.19) [7]. The preferential solvation parameter values obtained from the QLQC method can be compared with those obtained from the IKBI method. The differences between the $\delta x_{A(B)}$ obtained from the former and the δx_{AB} obtained from the latter, noted earlier, should, however, be remembered. Positive values of $\delta x_{A(B)}$ were obtained for aqueous 1,2-ethanediol, dimethylsulfoxide, formamide, and hexamethyl phosphoric triamide. These positive values signify stronger mutual than self-interactions of the components. Negative values of $\delta x_{A(B)}$ were obtained for aqueous acetonitrile, sulfolane, *N,N*-dimethylformamide, pyridine, triethylamine, acetone, tetrahydrofuran, dioxane, and the alkanols—methanol, ethanol, 1- and 2-propanol, *t*-butanol, and 2-butanol—signifying stronger self-interactions than mutual interactions of the components [22]. These results are in substantial agreement with those described in detail earlier, obtained from the IKBI method. In more recent publications, Batov et al. [22a] reported the use of the partial molar heats of mixing of the components for the calculation of the local mole fractions of the components in the solvation shells in binary aqueous mixtures. The cosolvents studied were methanol, ethanol, *n*- and *i*-propanol, *t*-butanol, tetrahydrofuran, 1,4-dioxane, acetone, acetonitrile, hexamethyl phosphoric triamide, and

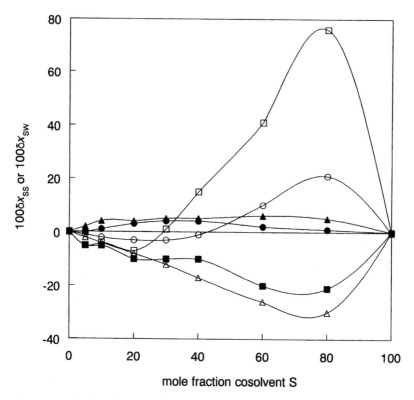

FIGURE 4.5 Preferential solvation parameters obtained by Batov et al. [22a] from partial molar heats of mixing, $100\delta x_{SS}$ and $100\delta x_{SW}$, of ethanol (● and ○), acetonitrile (▲ and △), and hexamethylphosphoramide (■ and □).

dimethylsulfoxide. Their results agree substantially with those reported here; see Fig. 4.5 for examples.

Kirkwood–Buff integrals have also been calculated for numerous binary nonaqueous solvent mixtures, and preferential solvation parameters have been derived from them in many cases by Marcus [22] and Matteoli [23], among others. Because the discrepancy of the molar volumes of nonaqueous solvents tends to be smaller than those existing when one of the solvents is water, the correction for the volume effects (i.e., from δx_{AA} to $\delta x'_{AA}$, etc.) is less important than for aqueous mixtures. Many of the systems that have been studied are listed in Table 4.2, where the self-preference of component A is denoted by the sign of δx_{AA}. Where this is designated by "near zero," the $\delta x_{AA}(x_B)$ curve is often S-shaped but with small values in either direction, $|\delta x_{AA}| \leq 0.02$. A positive sign of δx_{AA} signifies, as before, self-association of component A. The general trends are as for mixtures involving water. When one of the components (say, A) is highly polar and/or hydrogen bonding and the other is not, the former tends to self-aggregate, hence has positive values of the preferential solvation parameters δx_{AA}. If both components have similar polarities and/or hydrogen-bonding abilities, δx_{AA} values near zero result. Negative values of δx_{AA} are obtained when one of the components is a good electron pair or hydrogen bond donor and the other a good acceptor, e.g., as with chloroform and triethylamine. It is instructive to note that the results from the Kirkwood–Buff integrals are in accord with those from the QLQC method, Eq. (4.17). It is noteworthy that this is the case, although the latter is based on a rather simple model of the interactions whereas the former is based on rigorous considerations, up to the calculation of the correlation volume [22].

4.4   USE OF CHEMICAL PROBES

The use of a chemical probe as a solute in a binary mixture of solvents A and B for the exploration of preferential solvation poses the following problem. A certain property of the probe, say X, is determined spectroscopically in the solvent mixtures, and its dependence on the properties of the neat components, X_A and X_B, and the compositions of the mixtures is determined. In rare cases this dependence is linear with the solvent composition, but generally it is found to be nonlinear:

$$X = x_A X_A + x_B X_B + \Delta X \tag{4.33}$$

The deviations from linearity, ΔX, are generally ascribed to the preferential solvation of the probe by one component of the solvent mixture. For instance, the preferential solvation of the zwitterionic betaine solvatochromic indicator 4-(2,6-diphenylpyridino)-2,4,6-triphenylphenolate in binary solvent mixtures was investigated extensively. The quantity used was $E_T(30)/\text{kcal mol}^{-1} = 28,590/(\lambda_{max}/\text{nm})$, where $1\,\text{cal} = 4.184\,\text{J}$ and λ_{max} is the wavelength of the lowest energy peak in the UV–visible spectrum of the probe. It is generally found (as summarized in Tables 2.24 and 2.25 in Section 2.4) that $E_T(30)$ is not linear with the solvent composition, i.e., $\Delta E_T(30) \neq 0$ in the following Eq. (4.34):

$$E_T(30) = x_A E_T(30)_A + x_B E_T(30)_B + \Delta E_T(30) \tag{4.34}$$

TABLE 4.2   Nonaqueous Binary Mixtures Studied by Means of the Kirkwood–Buff Integrals and the Signs of the Resulting Self-Association Preferential Solvation Parameters, δx_{AA}

Solvent B	Solvent A				
	Methanol	Ethanol	Acetone	Triethylamine	Other
n-Hexane	(Immiscible)	Positive	Positive		[d,e,i]
c-Hexane	(Immiscible)	Positive	Positive		[c,g,i]
n-Heptane	(Immiscible)	Positive	Positive	Near zero	
Benzene	Positive	Positive	Negative	Negative	[h]
Toluene	Positive	Positive	Near zero		
Chloroform		Positive	Negative	Negative	
Tetrachloromethane	Positive[b]	Positive[b]			Positive[b]
Chlorobenzene	Positive	Negative[d]	Near zero	Negative	
Nitromethane	Positive	Positive	Near zero		
Acetonitrile	Positive	Positive			
Pyridine	Near zero	Near zero	Near zero		
Triethylamine		Near zero			
Sulfolane	Positive	Positive			
Dimethylsulfoxide	Negative	Near zero			
Formamide	Near zero		Positive		
NMF	Near zero[a]	Near zero[a]			
NMA	[j]	[j]			
DMF	Negative	Near zero[a]	Near zero		
DEF			Near zero		
THF	Near zero				
Dioxane			Positive		
Acetone			Near zero		
Methanol			Near zero[c]	Negative	Negative Positive[c]
Ethanol	Near zero[c]		Near zero[d]		
n-Propanol	Negative[d]	Near zero[d,f]		Near zero	
i-Propanol	Near zero[d,f]	Near zero[d,f]			
n-Butanol	Positive[d,f]				
t-Butanol	Negative[d,f]	Negative[d]		Near zero	
n-Decanol	Positive[i]				
Ethylene glycol		Positive[d]	Positive		

The data are from Ref. 22 unless otherwise noted.
[a] J Zielenkiewicz. J Chem Soc Faraday Trans 94:1713, 1998.
[b] Kato [11] (see Zaitsev et al. [17]), A = n-propanol.

Reference footnotes continue on pg. 169.

TABLE 4.2 Footnotes continued

---

[c]E Matteoli. J Phys Chem B101:9800, 1997, A = benzene for B = cyclohexane and A = n-decanol for B = methanol.
[d]Results from the QLQC method, Marcus [7].
[e]A = 1-butanol, B = n-hexane.
[f]Y Marcus. J Chem Soc Faraday Trans 87:1843, 1991.
[g]A = 2,3-dimethylbutane; MCA Donkersloot. J Solution Chem 8:293, 1979.
[h]Near zero for A = tetrachloromethane, positive for A = 1,2-dichloroethane; A Guha, D Mukherjee. J Indian Chem Soc 72:609, 1995.
[i]Positive for A = benzene, near zero for A = n-hexadecane; E Matteoli. J Phys Chem B101:9800, 1997.
[j]The value of δx_{BA} obtained for A = methanol and ethanol by J Zielkiewicz. Phys Chem Chem Phys 2:2925, 2000.

---

and $E_T(30)$ may even show an extremum. A quantitative approach due to Dawber et al. [24,25] plots the measured $E_T(30)$ of the mixture against the composition and notes the mole fraction y_A at which $E_T(30)(x_A) = y_A E_T(30)_A + (1 - y_A)E_T(30)_B$. The solvent sorting parameter p is then obtained from

$$p = y_A(1 - x_A)/(1 - y_A)x_A \tag{4.35}$$

A value $p = 1$ corresponds to random solvation of the probe by the two solvents, whereas $p > 1$ corresponds to preferential solvation by the more polar solvent A [i.e., when $E_T(30)_A > E_T(30)_B$] and vice versa for $p < 1$. In order to remove the effect of the absolute difference of the $E_T(30)$ values of the two solvents, Chatterjee and Bagchi [26,27] proposed to define the *local* mole fraction of solvent A around the probe, x_A^L, by using the quantity $\delta x_A = \Delta E_T(30)/[E_T(30)_A - E_T(30)_B] = x_A^L - x_A$. Naturally, if $\delta x_A > 0$ then solvent A preferentially solvates the probe.

Thus, the property X is said to be the locally weighted average of the properties of neat solvents A and B that, instead of being linear with the *nominal composition* of the solvent mixture, is linear with the *local composition* in the environment of the probe that is sensitive to the solvent composition:

$$X = x_{A(X)}^L X_A + x_{B(X)}^L X_B \tag{4.36}$$

Therefore, because X_A and X_B are known and, of course, $x_B^L(x) = 1 - x_{A(X)}{}^L$, the function $x_{A(X)}^L$ can be readily obtained from the determination of X as a function of x_A: $x_{A(X)}^L = (X(x_A) - X_B)/(X_A - X_B)$.

The problem that arises is that if a different probe and/or property Y is used, in general the local composition function $x^L_{A(Y)}$ calculated from an expression similar to (4.36) but with Y_A and Y_B instead of X_A and X_B is different from that obtained by the use of the probe/property X. If a different probe is used, then all that can be said is that the new probe is solvated differently from the previous probe molecule. However, if for a given probe different properties (e.g., two UV–visible absorption wavelengths, IR wavenumbers of different bonds, ^1H and ^{13}C NMR chemical shifts) yield different local mole fractions, as they are apt to do, what is then actually measured? What, then, is the "true" preferential solvation in the binary solvent mixture? In such cases there is no simple answer.

Nevertheless, it was found empirically that in favorable cases several probes/properties may still provide similar values of the local composition around them as a function of the nominal composition of binary solvent mixtures. The premise now is that it is of minor interest how a given chemical probe is preferentially solvated in the mixture but of major interest whether and how any other solute molecule (or atom or ion) will be preferentially solvated in it. In such cases it is useful to compare these values of $x_A{}^L(x_A)$ with those obtained from, say, the reversed Kirkwood–Buff integral (IKBI) or the quasi-lattice quasi-chemical (QLQC) method or from spectroscopic measurements on the solvent mixture in the absence of a solute used as probe. Such a comparison was made by Marcus [22], for instance, for binary solvent mixtures where microheterogeneity takes place, so that there are regions in the mixed solvent in which one or the other component predominates. In such systems, with large positive excess Gibbs energies of mixing, a measured solvatochromic parameter X may indeed be linear with the *local* mole fractions obtained from the IKBI and QLQC approaches and the $x^L_{A(X)}(x_A)$ curves are concave upward. For systems where component A was water and B was tetrahydrofuran, pyridine, or acetonitrile, the polarity/polarizability parameter π^* was linear with $x_A{}^L$ as in Eq. (4.36). In other systems, where the excess Gibbs energy is negative or positive but small (dimethylsulfoxide, formamide, and formic acid), a cooperative effect between the solvents appears to take place, producing a $\pi^*(x_A)$ curve with a maximum, which is not simply related to the local mole fraction of water.

The two-step models of Skwierczynski and Connors [28] and of Rosés, Bosch, et al. [29–37] were devised in order to deal with such a cooperative effect. The solvation of the probe P is described by the two replacement equilibria when cosolvent B is added to a solution of P in solvent A:

$$PA_2 + 2B \rightleftharpoons PB_2 + 2A \tag{4.37a}$$

and

$$PA_2 + B \rightleftharpoons PAB + A \tag{4.37b}$$

Solvation of the probe by the solvent adduct AB is thus assumed for the second step. (The stoichiometries involved in the two steps are arbitrary but were the simplest found to permit good fitting of the data.) Skwierc-zynski and Connors [28] presented the normalized Dimroth–Reichardt polarity parameters of solvent mixtures, $E_T^N = [E_T(30) - 30.7]/32.4$, in terms of the equilibrium constants K_1 and K_2 of these equilibria, and their values for aqueous cosolvent mixtures are shown in Table 4.3. Note that according to Skwierczynski and Connors [28] for aqueous mixtures of alkanols, alkanediols, and dimethylsulfoxide solvation of the probe by the adduct is not required for fitting the data, but this is necessary for other mixtures.

Rosés, Bosch, and coworkers [29–37] defined preferential solvation parameters $f_{B/A}$ and $f_{AB/A}$ as

$$f_{B/A} = (x_B^L/x_A^L)/(x_B/x_A) \tag{4.38a}$$

$$f_{AB/A} = (x_{AB}^L/x_A^L)/(x_B/x_A) \tag{4.38b}$$

[compare Eq. (4.38a) with Eq. (4.35)]. The solvatochromic or other measured property value X in the mixture is then

$$X = x_A^L X_A + x_B^L X_B + x_{AB}^L X_{AB} \tag{4.39}$$

[compare Eq. (4.36)] or, using the preferential solvation parameters,

$$X = X_A + [f_{B/A}(X_B - X_A)x_B^2 + f_{AB/A}(X_{AB} - X_A)x_B]/$$
$$[x_A^2 + f_{B/A}x_B^2 + f_{AB/A}x_Ax_B] \tag{4.40}$$

with three unknown parameters that have to be determined: $f_{B/A}$, $f_{AB/A}$, and X_{AB}. Values of these parameters for the normalized Dim-roth–Reichardt betaine probe, E_T^N are shown in Table 4.4. Some binary mixtures studied by Bosch, Rosés, et al. [29–37] can also be described with the single-step model, i.e., $f_{AB/A} = 0$ and $E_{T\,AB}^N = (E_{T\,A}^N + E_{T\,B}^N)/2$. The same results are obtained if $f_{AB/A} = f_{B/A} + 1$ and $E_{T\,AB}^N = (E_{T\,A}^N + f_{B/A} E_{T\,B}^N)/(1 + f_{B/A})$. In some of the cases reported by Bosch, Rosés, et al. [29–37] E_T^N for the binary solvent mixture was not

TABLE 4.3 Constants K_1 and K_2 for the Normalized $E_T(30)$ Solvatochromic Parameters, $E_T^N = [E_T(30) - 30.7]/32.4 = 1 - (1 - E_{T\ \text{cosolvent}}^N)[K_1 K_2 x^2 + K_1(1 + e^{-100 K_2})x(1 - x)/2]/[K_1\ K_2 x^2 + K_1 x(1 - x) + (1 - x)^2]$, of aqueous cosolvent mixtures at 298.15 K, where x is the mole fraction of the cosolvent

Cosolvent	Ref.	$E_{T\ \text{cosolvent}}^N$	K_1	K_2
Methanol	a	0.762	4.4	0
Ethanol	a	0.654	6.1	0
2-Propanol	a	0.546	6.4	0
2-Methyl-2-Propanol	b	0.389	205.1	0.43
1,2-Ethanediol	a	0.790	3.9	0
Tetrahydrofuran	a	0.207	22.3	0.17
Dioxane	a	0.164	8.8	0.28
Acetone	a	0.355	10.1	0.18
Pyridine	a	0.302	18.5	0.42
Acetonitrile	a	0.460	7.9	0.05
N, N-Dimethylformamide	a	0.386	8.8	0.74
N-Methylpyrrolidin-2-one	c	0.354	3.25	0
Dimethylsulfoxide	a	0.444	2.8	0

[a]RD Skwierczynski, KA Connors. J Chem Soc Perkin Trans 2 1994:467.
[b]H Langhals. Angew Chem Int Ed Engl 21:724, 1982.
[c]M Kemell, P Pirilä. J Solution Chem 92:87, 2000.

determined directly with the Dimroth–Reichardt betaine indicator but was back-calculated from the Kamlet–Taft solvatochromic parameters α, β, and π^*, known for such mixtures, and related to the $E_T(30)$ values in the same manner as for neat solvents [38]. Furthermore, in later publications these authors reported numerical values of their parameters different from those in earlier publications for the same binary mixtures, probe, and temperature. Hence, the reported preferential solvation parameters $f_{AB/A}$ and $f_{B/A}$ should be taken as fitting parameters for the E_T^N data rather than fundamental descriptors of the preferential solvation of the probe by the components of the solvent mixture. Still, $f_{B/A} > 1$ denotes a preference of the probe molecule for having component B in its near environment, whereas $f_{B/A} < 1$ denotes a preference for component A.

In any case, it needs to be stressed again that the preferential solvation described by Eq. (4.40) pertains to the *preferential solvation of the solvatochromic probe* and does not say anything concerning the mutual or self-solvation of the molecules of the solvents in the absence of the probe

TABLE 4.4   Values of the Selective Solvation Parameters $f_{B/A}$ and $f_{AB/A}$ and the Normalized Dimroth–Reichardt Solvatochromic Parameter for the Adduct AB, $E^N_{T AB}$, in Mixtures of Solvents A and B at 298.15 K

Solvent A	$E^N_{T A}$	Solvent B	$E^N_{T B}$	$f_{B/A}$	$f_{AB/A}$	$E^N_{T AB}$
Water	1.000	Methanol[a]	0.764	4.4	0	—
		Ethanol[a]	0.652	6.1	0	—
		1-Propanol[a]	0.617	12.2	0	—
		2-Propanol[a]	0.543	6.4	0	—
		1,2-Ethanediol[a]	0.784	3.9	0	—
		1,2-Propanediol[a]	0.722	7.1	0	—
		1,3-Propanediol[a]	0.747	7.3	0	—
		Tetrahydrofuran[a]	0.207	3.8	22.3	0.604
		1,4-Dioxane[a]	0.164	2.5	8.8	0.582
		Acetone[a]	0.352	1.8	10.1	0.676
		Acetonitrile[a]	0.432	0.40	7.9	0.716
		Piperidine[a]	0.148	8.8	20.1	0.574
		Pyridine[a]	0.296	7.8	18.5	0.648
		2-Methylpyridine[a]	0.235	13.0	31.0	0.618
		2,6-Dimethylpyridine[a]	0.185	16.9	49.6	0.593
		Formamide[b]	0.775	0.31	3.0	0.794
		N-Methylformamide[b]	0.721	31	19	0.882
		Dimethylformamide[b]	0.396	6.1	7.9	0.702
		Dimethylsulfoxide[a]	0.432	2.8	0	—
Methanol	0.764	water[f]	1.000	0.26	1.3	0.817
		Ethanol[f]	0.652	1.0	2.0	0.705
		2-Butanol[f]	0.510	0.88	1.9	0.649
		2-Methyl-2-butanol[f]	0.321	0.06	1.0	0.483
		1,2-Ethanediol[f]	0.786	1.0	2.0	0.772
		2-Methoxyethanol[f]	0.656	1.5	2.5	0.698
		Acetonitrile[f]	0.468	333	7000	0.777
		Tetrahydrofuran[f]	0.212	0.74	5.1	0.686
		Chloroform[f]	0.315	1.6	5.7	0.671
		Dichloromethane[f]	0.313	0.06	1.9	0.606
		Benzene[f]	0.123	0.05	2.2	0.564
		Toluene[f]	0.121	0.08	2.9	0.573
		Dimethylsulfoxide[f]	0.443	50	110	0.763
		Pyridine[f]	0.294	0.03	0.72	0.425
		Formamide[b]	0.775	1.8	5.6	0.806
		N-Methylformamide[b]	0.721	3.7	4.8	0.798
		Dimethylformamide[b]	0.396	0.012	0.40	0.465
Ethanol	0.652	water[f]	1.000	0.14	1.1	0.706

TABLE 4.4 Continued

Solvent A	E_{TA}^N	Solvent B	E_{TB}^N	$f_{B/A}$	$f_{AB/A}$	E_{TAB}^N
Ethanol	0.652	Methanol[f]	0.764	1.0	2.0	0.705
		2,2,2-Trifluoroethanol[f]	0.897	0.53	1.5	0.570
		2-Methyl-2-Butanol[f]	0.321	1.6	2.6	0.734
		1,2-Ethanediol[f]	0.786	1.0	2.0	0.659
		2-Methoxyethanol[f]	0.656	2.4	3.4	0.825
		Acetone[f]	0.365	1.0	9.0	0.663
		Chloroform[f]	0.267	0.08	1.5	0.460
		Acetonitrile[f]	0.466	0.4	8.1	0.703
		Dimethylsulfoxide[f]	0.432	4	7.4	0.687
1-Propanol	0.614	Water[f]	1.000	0.10	2.7	0.673
		Acetone[f]	0.364	0.43	2.8	0.682
		Chloroform[f]	0.318	1.2	3.8	0.542
2-Propanol	0.543	Water[f]	1.000	0.34	4.0	0.620
		Acetonitrile[f]	0.466	0.38	7.5	0.630
		Dimethylsulfoxide[f]	0.432	2.0	3.1	0.614
		Formamide[b]	0.775	3.9	6.9	0.575
		N-Methylformamide[b]	0.721	351	69	0.586
		Dimethylformamide[b]	0.396	5.6	21	0.550
2-Methyl-2-Propanol	0.406	Methanol[c]	0.764	19	18	0.490
		Ethanol[c]	0.652	4.6	6.8	0.513
		2-Propanol[c]	0.543	1.9	3.7	0.487
		Hexane[c]	0.005	0.46	2.1	0.300
		Benzene[c]	0.123	0.007	1.1	0.215
		Formamide[b]	0.775	24	21	0.550
		N-Methylformamide[b]	0.721	749	196	0.502
		Dimethylformamide[b]	0.396	0.81	3.4	0.496
1,2-Ethanediol[f]	0.786	Water[f]	1.000	0.32	1.3	0.842
		Methanol[f]	0.759	1.0	2.0	0.772
		Ethanol[f]	0.659	0.61	1.6	0.753
		2-Methoxyethanol[f]	0.666	1.0	2.0	0.727
Acetone	0.355	Ethanol[d]	0.652	1.0	9.0	0.663
		2-Propanol[d]	0.543	2.4	6.5	0.682
		Trimethylphosphate[d]	0.398	1.2	2.0	0.398
		Dichloromethane[d]	0.324	2.1	7.4	0.389
		Chloroform[d]	0.272	3.3	5.8	0.448
		Tetrachloromethane[d]	0.056	0.10	1.2	0.235
		Cyclohexane[d]	0.006	0.15	2.7	0.267
		2-Methyl-2-Propanol[d]	0.406	0.96	2.7	0.448

TABLE 4.4 Continued

Solvent A	E_{TA}^N	Solvent B	E_{TB}^N	$f_{B/A}$	$f_{AB/A}$	E_{TAB}^N
Acetonitrile	0.462	Methanol[d]	0.764	0.003	20	0.777
		Ethanol[d]	0.652	2.5	20	0.703
		2-Propanol[d]	0.543	2.6	20	0.630
		2-Methyl-2-Propanol[d]	0.406	7.0	22	0.558
		Dimethylsulfoxide[d]	0.443	43	254	0.450
		Benzene[d]	0.123	1.3	4.7	0.419
Nitromethane	0.480	Methanol[e]*	0.766	2.1	30	0.778
		2-Propanol[e]	0.543	3.3	29	0.609
		2-Methyl-2-Propanol[e]	0.406	7.9	25	0.419
N-Methyl-formamide	0.721	Formamide[b]	0.775	0.13	1.1	0.767
Dimethylforma-mide	0.369	Formamide[b]	0.775	5.9	8.9	0.607
		N-Methylformamide[b]	0.721	125	40	0.571
Dimethylsulfoxide	0.443	Water[d]	1.000	0.57	2.0	0.542
		Methanol[d]	0.764	0.02	2.1	0.763
		Ethanol[d]	0.652	0.25	1.8	0.687
		2-Propanol[d]	0.543	0.51	1.6	0.614
		2-Methyl-2-Propanol[d]	0.406	0.72	1.6	0.548
		Chloroform[d]	0.319	0.19	0.31	0.473
		Tetrachloromethane[d]	0.056	0.01	0.91	0.327
		Benzene[d]	0.123	0.38	3.6	0.393
		Acetonitrile[d]	0.462	0.02	5.9	0.450
		Acetone[d]	0.364	1.04	2.8	0.448
		Formamide[g]	0.789	1.84	0	—
Benzene	0.123	Pyridine[f]	0.304	1.9	2.9	0.241
		Nitrobenzene[f]	0.329	5.7	6.7	0.298
Tetrachloro-methane	0.056	Chloroform[f]	0.277	23	22	0.159
		Trimethylphosphate[g]	0.380	5.49	0	—
Tributylphosphate	0.27	Trimethylphosphate[g]	0.380	1.06	0	—

[a]RD Skwierczynski, KA Connors. J Chem Soc Perkin Trans. 2 1994:467; where E_{TAB}^N is given, it is the mean between the E_T^N of solvents A and B.
[b]K Herodes, I Leito, I Koppel, M Rosés. J Phys Org Chem 12:109, 1999.
[c]E Bosch, F Rived, M Rosés, J Chem Soc Perkin Trans 2 1996:2177.
[d]M Rosés, C Rafols, J Ortega, E Bosch, J Chem Soc Perkin Trans 2 1995:1607.
[e]E Bosch, M Rosés, K Herodes, Il Koppel, I Leito, Iv Koppel, V Taal, J Phys Org Chem 9:403 1996.* At 303.15 K, E_{TA}^N where = 0.493.
[f]J Ortega, C Rafols, E Bosch, M Rosés. J Chem Soc Perkin Trans 2 1996:1497.
[g]E Bosch, M Rosés. J Chem Soc Faraday Trans 88:3541, 1992.

or concerning the preferential solvation of other kinds of probes or solutes in general (see Section 5.1). It is, therefore, natural that different probes in a given binary solvent mixture have quite different values of $f_{B/A}$ and $f_{AB/A}$ [29–37], denoting different preferences regarding the components of the mixture.

4.5 DIRECT SPECTROSCOPIC AND STRUCTURAL STUDIES

In cases in which the one or both of the solvents in a binary mixture have suitable spectroscopic properties, these can be utilized in order to ascertain the preferential solvation, if any, in the mixture, without the need for a probe. Only few studies have been made in this direction, although the structure of solvent mixtures, without reference to preferential solvation, has been studied extensively by spectroscopic methods; see Section 3.3. In general, only dilute solutions of one solvent in the other have been made, but in some cases much or all of the composition range was covered.

The infrared spectra of dilute solutions of acetonitrile, dimethylsulfoxide, and acetone in a large number of organic solvents were studied by Fawcett et al. [39–44], and the solvent-induced frequency shifts (SIFSs) were recorded. The self-interactions (association to dimers, etc.) of the former three solvents as solutes and their mutual interactions with the other solvents were explained in terms of the Lewis acid–base concept and, where required, also by invoking nonspecific interactions. The SIFSs of the v_2 vibration of acetonitrile in its solutions were in both directions because acetonitrile can both receive hydrogen bonds from and donate hydrogen bonds to (through the methyl hydrogen atoms) the other solvent. The data fitted the expression $\Delta v_2 = -3.7 + 0.32DN - 0.13AN \, \text{cm}^{-1}$, where DN and AN are the donor and acceptor numbers of the solvent [38]. Formamide and N-methylformamide as solvents for the acetonitrile are outliers in this correlation because they act as both Lewis acid and Lewis base. Acetone and dimethylsulfoxide act only as Lewis bases, and their acceptance of hydrogen bonds from protic solvents competes with their self-association (through dipole interactions) in aprotic solvents [39]. The SIFSs are for acetone $\Delta v_3 = 1.3 - 0.29AN \, \text{cm}^{-1}$ and for dimethylsulfoxide $\Delta v(S{=}O) = 22.0 - 1.38AN \, \text{cm}^{-1}$. In some cases higher concentrations of one solvent in the mixture—acetone in tetrachloromethane and in nitrobenzene and ethylene carbonate (EC) in water—were studied in order to obtain a quantitative measure of the self-association [40–41]. The dimerization constant K_d of acetone was obtained from plots of the concentration of acetone against the square root of the integrated intensity, A_d, of the resolved dimer band in the spectrum:

$$c(Me_2CO) = A_d^{1/2}/a_d^{1/2}K_d^{1/2} + 2A_d/a_d \qquad (4.41)$$

where a_d is the molar absorbance of the dimer. For the acetone dimers, which could be linear and/or cyclic, the values $K_d = 1.1$ and 0.4 L mol^{-1} were obtained in tetrachloromethane and nitrobenzene, respectively. In the case of aqueous EC, both the water stretching mode infrared band at 2700 to 3900 cm^{-1}, studied by attenuated total reflection, and the EC bands between 1000 and 2050 cm^{-1} were measured. These provided information on the effect of the EC on the water structure on the one hand and the hydrogen bonding and hydrophobic hydration of EC in this system on the other, but on the whole, little preferential solvation could be deduced from the results.

The microheterogeneity noted in mixtures of water and acetonitrile from thermodynamic and other data (see earlier for results from the IKBI method) was confirmed by infrared spectral measurements by Bertie and Lan [45] of both the C≡N stretching vibration in the acetonitrile and the O−H one in the water. It was confirmed as well by the O−D stretching vibration in 20% deuterated water measured by Takamuku et al. [46]. In very dilute acetonitrile ($x_{MeCN} \leq 0.05$) there are more hydrogen bonds than in pure water because more H ⋯ N bonds are formed than H ⋯ O are destroyed [45], or, as in the case of other partly hydrophobic solutes, the H-bonded water structure is enhanced [46]. The microheterogeneity range is $0.3 \leq x_{MeCN} \leq 0.5$ [45] or $0.2 \leq x_{MeCN} \leq 0.6$ [46]; acetonitrile clusters certainly exist but water clusters may [46] or may not [45] exist, and in the latter case linear chains or hexagons of water are likely to occur. The IR results of Takamuku et al. [46] also confirmed their X-ray diffraction data, obtained for 10 compositions of the water and acetonitrile mixtures over the entire composition range; see also Section 3.3.

An example of the use of NMR chemical shift measurements for the direct study of the preferential solvation in binary solvent mixtures was provided by Muller [47]. He used the ^{19}F chemical shifts of the trifluoromethyl group in perfluorohexane + hexane mixtures at 25 to 55°C, the lower temperature being just above the upper consolute temperature of this mixture. These shifts were interpreted in a manner implicitly related to the QLQC method, using $(2/Z)\Delta e_{AB}/k_B T$ of Eq. (4.16) as the fitting parameter and volume fractions rather than mole fractions for weighting the observed shifts. The self-interactions of both components were found to be preferred over the mutual interactions. The lattice parameter Z

increased from 7 to 9 with increasing temperatures but Δe_{AB} decreased although relatively less.

Most other studies of binary solvent mixtures by diffraction, spectroscopic, and computer simulation methods (Chapter 3) did not specifically address themselves to obtaining the extent of preferential solvation that takes place in the mixtures.

REFERENCES

1. GM Wilson. J Am Chem Soc 86:127, 1964.
2. H Renon, JM Prausnitz. AIChE J 14:135, 1968.
3. Y Marcus, Aust J Chem 36:1719, 1983.
4. EA Guggenheim. Proc R. Soc A169:134, 1938.
5. DS Abrams, JM Prausnitz. AIChE J 21:116, 1975.
6. K Hagemark. J Phys Chem 72:2316, 1968.
7. Y Marcus. J Chem Soc Faraday Trans 185:381, 1989, corrected in Y Marcus, Y Migron. J Phys Chem 95:400, 1991.
8. JG Kirkwood, FP Buff. J Chem Phys 19:774, 1951.
9. A Ben-Naim. J Chem Phys 67:4884, 1977.
10. E Matteoli, L Lepori. J Chem Phys 80:2856, 1984.
11. T Kato. J Phys Chem 88:1248, 1984.
12. A Ben-Naim. J Phys Chem 93:3809, 1989; Pure Appl Chem 62:25, 1990.
13. J-L Kim. Z Phys Chem NF 113:129, 1978.
14. Y Marcus. J Chem Soc Faraday Trans 185:3019, 1989.
15. E Matteoli. J Phys Chem B101:9800, 1997.
16. AL Zaitsev, YuM Kessler, VE Petrenko. Zh Fiz Khim 59:2728, 1985.
17. AL Zaitsev, VE Petrenko, YuM Kessler. J Solution Chem 18:115, 1989.
18. MJ Blandamer, NJ Blundell, J Burgess, HJ Cowles, IM Horn. J Chem Soc Faraday Trans 186:277, 1990.
19. MJ Blandamer, NJ Blundell, J Burgess, HJ Cowles, IM Horn. J Chem Soc Faraday Trans 186:283, 1990.
20. MJ Blandamer, J Burgess, A Cooney, HJ Cowles, IM Horn, KJ Martin, KW Morcom, P Warrick Jr. J Chem Soc Faraday Trans 186:2209, 1990.
21. Y Marcus. Monatsh Chem 132:1387, 2001.
22. Y Marcus. J Chem Soc Faraday Trans 87:1843, 1991.
22a. DV Batov. Zh Obshch Khim 68:204, 1998; DV Batov, AM Zaichikov, VP Slyusar, VP Korolev. Zh Obshch Khim 69:19523, 1999.
23. E Matteoli. J Mol Liq 79:101, 1999.
24. JG Dawber, J Ward, RA Williams. J Chem Soc Faraday Trans 184:713, 1988.
25. JG Dawber. J Chem Soc Faraday Trans 186:287, 1990.
26. P Chatterjee, S Bagchi. J Chem Soc Faraday Trans. 18:1785, 1990.
27. P Chatterjee, S. Bagchi. J Chem Soc Faraday Trans 187:587, 1991.

28. RD Skwierczynski, KA Connors. J Chem Soc Perkin Trans 2 1994:467.
29. E Bosch, M Rosés. J Chem Soc Faraday Trans 88:3541, 1992.
30. M Rosés, C Rafols, J Ortega, E. Bosch. J Chem Soc Perkin Trans 2 1995:1607.
31. E Bosch, M Rosés, K Herodes, I Koppel, I Leito, I Koppel, V Taal. J Phys Org Chem 9:403, 1996.
32. J Ortega, C Rafols, E Bosch, M Rosés. J Chem Soc Perkin Trans 2 1996:1497.
33. E Bosch, F Rivet, M Rosés. J Chem Soc Perkin Trans 2, 1996:2177.
34. C Ràfols, M Rosés, E Bosch. J Chem Soc Perkin Trans 2, 1997:243.
35. U Buhvestov, F Rived, C Ràfols, E Bosch, M Rosés. J Phys Org Chem 11:185, 1998.
36. M Rosés, U Buhvestov, C Ràfols, F Rived, E Bosch. J Chem Soc Perkin Trans 2, 1997:1341.
37. K Herodes, I Leito, I Koppel, M. Rosés. J Phys Org Chem 12:109, 1999.
38. Y Marcus. Chem Soc Rev 22:409, 1994.
39. WR Fawcett, GJ Liu, TE Kessler. J Phys Chem 97:9293, 1993.
40. WR Fawcett, AA Kloss. J Phys Chem 100:2019, 1996.
41. WR Fawcett, AA Kloss. J Chem Soc Faraday Trans 92:3333, 1996.
42. DK Cha, AA Kloss, AC Tikanen, WR Fawcett. Phys Chem Chem Phys 1:4785, 1999.
43. PA Brooksby, WR Fawcett. J Phys Chem A101:8307, 2000.
44. PA Brooksby, WR Fawcett. Spectrochim Acta A57:1207, 2001.
45. JE Bertie, Z Lan. J Phys Chem B101, 4111, 1997.
46. T Takamuku, M Tabata, A Yamaguchi, J Nishimoto, M Kumamoto, H: Wakita, T Yamaguchi. J Phys Chem B102:8880, 1998.
47. N Muller. J Phys Chem 83:1393, 1979.

5

Preferential Solvation of Solutes

5.1 GENERAL CONSIDERATIONS

As seen in Section 4.4, the use of chemical probes in binary solvent mixtures can, in the optimal case, provide information on the solvation properties of the mixture: polarity, Lewis basicity, and Lewis acidity. This depends on the convergence of the results for several probes with different functional groups relevant to the property measured so that the probes can represent a "general solute." In other cases, the chemical probes provide information on the preferential solvation of the probe molecule itself by the components of the solvent mixture. However, this information cannot be generalized to other solutes and is of limited interest. The question now arises: are there means to determine the preferential solvation in the solvent mixture of a given desired solute (molecule or ion) that cannot act as a chemical probe itself?

One answer is that the solute may still have some spectroscopic properties that are sensitive to the composition of its immediate environment, i.e., the solvation shells around the solute particle. Else, the spectroscopic properties of one or the other (or both) of the solvents may be sensitive to the presence of the solute in question. These properties can then be employed for the determination of the preferential solvation of the solute. In other cases, thermodynamic data may be used for the same purpose. The inverse Kirkwood–Buff integral (IKBI) and the quasi-lattice quasi-chemical (QLQC) methods, among others, can be used in ternary systems involving a (not necessarily dilute) solution of the solute in the binary solvent mixture. The general approach is similar to that used for these methods in the binary mixtures (see Section 4.3), but necessary modifications have to be introduced.

Consider a solute particle S in a binary solvent mixture, A + B, at high dilution of the solute. The solute is solvated, in principle, by both components but may be preferentially solvated by one of them. In such a case, the *local mole fraction*, x_{AS}^L, which describes the mole fraction of component A in the surroundings of the particle of the solute S, is different from the bulk mole fraction of A, x_A, but in any case:

$$x_{AS}^L + x_{BS}^L = x_A + x_B = 1 \tag{5.1}$$

One way to obtain the local mole fractions of the solvents around the solute in a binary solvent mixture is to use spectroscopic properties of the solute. The quantity x_{AS}^L has been *defined* as

$$x_{AS}^L = (v_{(AB)S} - v_{BS})/(v_{AS} - v_{BS}) \tag{5.2a}$$

or

$$x_{AS}^L = (\delta_{(AB)S} - \delta_{BS})/(\delta_{AS} - \delta_{BS}) \tag{5.2b}$$

where v is the frequency (wavenumber) of absorption, generally of the lowest energy absorption band, and δ is the nuclear magnetic resonance (NMR) chemical shift. Similar expressions can be written for other spectroscopic quantities. The subscript (AB)S pertains to the solute S in the solvent mixture and AS and BS pertain to the solute in the neat solvents. It is assumed in these definitions that the light absorbance frequency shift depends linearly on the proportion of the given solvents in the solvation shell of the solute:

$$v_{(AB)S} = x_{AS}^L v_{AS} + x_{BS}^L v_{BS} \tag{5.3a}$$

and similarly for the NMR chemical shift:

$$\delta_{(AB)S} = x_{AS}^L \delta_{AS} + x_{BS}^L \delta_{BS} \tag{5.3b}$$

The preferential solvation can also be specified in terms of the *preferential solvation parameter*, δx_{AS}, which describes the excess or deficiency of molecules of component A in the vicinity of S referred to the bulk composition of the solvent mixture:

$$\delta x_{AS} = x_{AS}^{L} - x_A \tag{5.4}$$

Positive values of δx_{AS} signify that solvent A is preferred in the vicinity of solute S, whereas negative values denote preference for the solvent B. Another way to express these preferences is by means of the position of the isosolvation point, x_A^{iso}, which is the bulk solvent composition at which the frequency v or the chemical shift δ in the mixture lies midway between those of the neat solvents. Then $x_A^{iso} < 0.5$ denotes preferential solvation by solvent A and $x_A^{iso} > 0.5$ denotes preferential solvation by B. A still further way to express the preferential solvation is by means of the *preferential solvation constant* $K_{AB/S}$ (instead of the quantity p employed for the binary solvent mixture itself, Section 4.1):

$$K_{AB/S} = [x_{AS}^{L}/(1 - x_{AS}^{L})]/[x_A/(1 - x_A)] \tag{5.5}$$

Values of $K_{AB/S} > 1$ denote preferential solvation of the solute S by solvent A and values of $K_{AB/S} < 1$ denote preferential solvation by B. No significant preferential solvation takes place when $K_{AB/S} \approx 1$. If the solvent–solvent interaction energies in the solvation shell are the same as in the bulk, then the preferential solvation constant $K_{AB/S}$ depends on the difference of the solute–solvent interaction energies e only: $K_{AB/S} = \exp[(e_{AS} - e_{BS})/k_B T] = \exp[\Delta_t G°(S,A \rightarrow B)/RT]$, where $\Delta_t G°$ is the standard molar Gibbs energy of transfer of the solute S from solvent A to solvent B. Because this condition is unlikely to occur, $K_{AB/S}$ will in general depend on the solvent–solvent interactions too.

The problem with these approaches to the quantitative description of the preferential solvation of the solute S in the solvent mixture A + B is that for a given solute different spectroscopic properties do not necessarily provide the same values of x_{AS}^{L} (hence of δx_{AS}, x_A^{iso}, or $K_{AB/S}$). This arises from the breakdown of the assumption that the shifts in v or in δ or in similar measures are proportional to the fractional composition of the solvents in the solvation shell of the solute in the mixture. Other ways to obtain the local mole fractions have, therefore, been sought.

The preferential solvation of solute A in the solvent mixture A + B can also be described by the stepwise replacement of, say, solvent A from the solvate SA_N, where N is the solvation number, by solvent B:

$$SA_N + B \rightleftharpoons SA_{N-1}B + A \qquad K_{A/B1} \qquad (5.6a)$$

$$\vdots \qquad\qquad \vdots$$

$$SA_nB_{N-n} + B \rightleftharpoons SA_{n-1}B_{N+1-n} + A \qquad K_{A/BN+1-n} \qquad (5.6b)$$

$$\vdots \qquad\qquad \vdots$$

$$SAB_{N-1} + B \rightleftharpoons SB_N + A \qquad K_{A/BN} \qquad (5.6c)$$

where the $K_{A/Bi}$ are the equilibrium quotients (independent of the concentration scale employed) for the stepwise replacements producing the solvates with i B solvent molecules in the solvation shell. It is assumed here that N remains constant and does not depend on the replacement step. An overall replacement constant can also be defined for the overall replacement equilibrium:

$$SA_N + NB \rightleftharpoons SB_N + NA \qquad K_N = \prod_{i=1}^{i=N} K_{A/Bi} \qquad (5.7)$$

These equilibria and equilibrium quotients can be handled in the manner familiar from coordination chemistry in solution [1,2]. The pitfall of the assumption that the shifts in v or in δ or in similar measures are proportional to the fractional composition of the solvents in the solvation shell of the solute should be avoided. It has been shown [3–8] that $K_{AB/S}$ defined in Eq. (5.5) may equal $K_N^{1/N}$ under certain conditions, including the condition that all the stepwise equilibrium quotients $K_{A/Bi}$ are the same. This is an unlikely condition because even if the solute–solvent interaction energies in the generalized equilibrium, Eq. (5.6b), are independent of the composition of the solvation shell, a statistical dependence of $K_{A/Bi}$ on i is to be expected.

5.2 NONELECTROLYTE SOLUTES

5.2.1 Spectroscopic Studies

Solvatochromic probes have been subjected in numerous reports to studies of their preferential solvation in solvent mixtures, expressed as in Eqs. (5.2) to (5.5). As mentioned earlier, this supplies information concerning the probes themselves that is of only limited interest but can be interpreted in terms of the interactions of these solutes with the component solvents. Bagchi and coworkers [3–8] studied the preferential solvation of N-alkyl-4-cyanopyridinium iodide in solvent mixtures by means of its charge transfer band [using the energy E rather than the wavenumber v in Eq. (5.2a)] and obtained values of the preferential solvation constant $K_{PS} = K_{AB/S}$ of Eq. (5.5). In mixtures of an alkanol with nitriles the alkanol was preferred near this ion-paired solute. In mixtures of aprotic dipolar solvents or polar solvents mixed with benzene the more polar solvent (say A) is preferred, the maximum in δx_{AS} tending toward B-rich compositions. In an extension of this work they studied the thermochromism of both N-alkyl-4-cyanopyridinium iodide and the Dimroth–Reichardt betaine in terms of the preferential solvation expressed as in Eqs. (5.2) to (5.5).

Merocyanine dyes were employed by Fayed and Etaiw [9] by Soroka and Soroka [10], among others, as probes in binary solvent mixtures. In aqueous mixtures of formamide, ethanol, i-propanol, DMF, acetonitrile, or dioxane the probe was preferentially solvated by the organic component. In mixtures containing methanol and the other solvents the former was preferred, except in mixtures with ethanol, where no preferential solvation was observed [9]. Similar approaches have been adopted by many authors, among whom were Phillips and Brennecke [11], who measured the frequency shift of the light absorption of phenol blue in mixtures of aprotic solvents. In all the cases studied the polar solvent was preferred. Taha et al. [12] measured the spectral shifts of the (neutral) cis-dicyano(1,10-phenanthrolino)iron(II) complex in 20 aqueous and non-aqueous cosolvent mixtures and calculated the local mole fractions of the solvent component B, x_B^L, according to Eq. (5.2a), the isosolvation composition, x_B^{iso}, and the solvent replacement equilibrium constant, $K_{AB/S}$, according to Eq. (5.5). Preferences of either component were noted, according to their chemical natures (Lewis acidities and polarities); water was preferred over dioxane, acetone, and acetonitrile but not over alkanols, DMF, and DMSO. Methanol was preferred over THF and acetonitrile, chloroform was preferred over aprotic dipolar solvents, but alkanols were preferred over chloroform, etc.

Acree and coworkers [13–17] studied probes that have a distinctive fluorescence, e.g., polynuclear aromatic hydrocarbons. They considered first cases in which only one component of the binary solvent mixture solvates the probe but in later studies allowed both components to solvate it. They then applied Eq. (5.3a) to solvent mixtures where one of the components self-associates. They also interpreted results in terms of the stepwise competitive exchange model, Eqs. (5.6) and (5.7).

These are just a few examples of the many studies of the preferential solvation of nonelectrolyte solutes obtained by spectroscopic measurements, utilizing essentially Eqs. (5.2) for obtaining the local mole fractions of the solvents in the solvation shell of the solute. Other approaches to the use of spectroscopic measurements are discussed further subsequently. A partial explanation of the results obtained for dipolar solutes in mixtures of solvents of different polarity was presented by Suppan [18,19] in terms of dielectric enrichment. The more polar solvent is enriched in the solvation shell and the preferential solvation index Z, Eq. (5.8), is essentially $\ln K_{AB/S}$ of Eq. (5.5). The index Z is given by

$$Z = C[f(\varepsilon_A) - f(\varepsilon_B)] V \mu_S^2 a_S^{-6} / \varepsilon_0 R T \qquad (5.8)$$

where solvent A is polar and B is not, C is a numerical constant (theoretically, $C = 3\pi^2/2$), $f(\varepsilon) = 2(\varepsilon - 1)/(2\varepsilon + 1)$, V is the mean molar volume of the solvent mixture, μ_S is the dipole moment of the solute, a_S is its radius, and ε_0 is the permittivity of vacuum. This approach needs modification if one of the solvents (or the solute) is protic and hydrogen bonds are formed. [Note: Eq. (5.8) has been modified from Suppan's [18] expression to make it dimensionally correct.]

The stepwise solvent replacement approach was used by Symons et al. [20–25] for the characterization by means of NMR and IR spectroscopy of the preferential solvation of various solutes in mixtures of protic and aprotic dipolar solvents. The probes included triethylphosphine oxide, trimethyl phosphate, dimethyl methylphosphonate, hexamethylphosphoric triamide [^{31}P NMR and ν(P=O) IR], acetone, esters [^{13}C NMR and ν(C=O) IR], acetonitrile [^{13}C and ^{14}N NMR and ν(C≡N) IR], and amides [ν(C=O) IR). In some cases, it was possible to establish the formation of tri-, di-, and monosolvates, but in many other cases, as with amide probes, only the di-, mono-, and non–hydrogen-bonded solvates were considered. The total solvation number $N = 2$ was then assumed, and the average replacement equilibrium constant $K^* = (K_{A/B1} K_{A/B2})^{1/2}$ [see Eq. (5.6)) was used to characterize the preferential solvation, Table 5.1. In all the cases, except for mixtures of methanol with dimethylsulfoxide, it was the protic solvent (methanol or water) that was preferred near the amide (N-methylformamide, N,N-dimethylformamide, N-methylacetamide, and N,N-dimethylacetamide) rather than the aprotic solvent (acetonitrile or tetrahydrofuran). Water was preferred over methanol, as could be expected from the hydrogen bond donation abilities of the solvents. The preference of strong hydrogen bond acceptor dimethylsulfoxide over methanol near the protic, singly substituted, amides is also readily understandable. The DMSO also preferentialy solvated acetic acid, methyl acetate, and acetonitrile in aqueous DMSO mixtures, but water

TABLE 5.1   Average Solvent Replacement Constant K^* of Amides NMF, DMF, NMA, and DMA[a], in Mixtures of Protic (A) and Aprotic (B) Solvents

| Solvent A | MeOH | MeOH | MeOH | H$_2$O | H$_2$O | H$_2$O | H$_2$O |
Solvent B	MeCN	DMSO	THF	MeOH	MeCN	DMSO	THF
NMF	0.40	1.40	0.50		0.45	0.75	0.40
DMF	0.25				0.50		
NMA	0.30	1.25	0.55	0.35	0.15	0.70	0.12
DMA	0.25	1.15	0.40	0.25	0.20	0.85	0.60

[a]Respectively N-methylformamide, N,N-dimethylformamide, N-methylacetamide, and N,N-dimethylacetamide.
Source: Refs. 20–25.

was preferred for the solvation of formamide and dimethylformamide and dimethylacetamide.

A further approach that should be mentioned is applicable where definite donor–acceptor bonding is known to take place between the solute and principally one component of the mixed solvent. This has been called by Nagy et al. [26] the competitive preferential solvation method (COPS) and involves the following postulates. The solute S is supposed to be solvated by both components, $i = $ A and B, of the solvent mixture according to their electronic–geometric affinities κ_{iS}. Hence, "inert" solvents and "free" solute do not exist in the solutions. Size differences are taken care of by the values of κ_{iS} that are independent of composition, pressure, and temperature. The molecules in the solvation shell are either "complexing," if their orbitals are in an orientation favorable to overlap with those of the solute, or "solvating," if they are not, but relax continuously between these states and the bulk solvent mixture. Therefore, κ_{iS} represents the total (complexing and solvating) affinity. The composition of the solvation shell is determined by the products of the affinities and the concentrations of the solvents: $\kappa_{iS}c_i$. The solute is now considered as partitioning among the solvent components, constituting a *partitioning in the homogeneous solution*:

$$c_S = \chi_{SA} + \chi_{SB} \qquad (5.9)$$

where

$$\chi_{Si} = c_S \kappa_{iS} c_i / \Sigma \, \kappa_{jS} c_j = c_S D_{Si} \qquad (5.10)$$

($i = $ A or B, $j = $ A and B), D_{Si} being a generalized (homogeneous) distribution factor. The solvent effect on any physicochemical property X of the solvated solute is considered to be additive in the solvation shell (of ill-defined extension):

$$X = D_{SA}X_{SA} + D_{SB}X_{SB} \tag{5.11}$$

where X_{Si} is the value of the property in neat solvent i.

The COPS method has been applied to situations where the solute S is an electron acceptor and one of the solvent components, say B, is the electron donor, whereas the other component, A, is a diluent that is not involved in donor–acceptor interactions. For example, if fast exchange takes place, the NMR chemical shift can be used as the property X:

$$\delta = D_{SA}\delta_{SA} + D_{SB}\delta_{SB} = \delta_{SB} + (\delta_{SB} - \delta_{SA})D_{SB} \tag{5.12}$$

because $D_{SA} = 1 - D_{SB}$. From Eq. (5.10) and using the (partial) molar volumes V_i (in L mol^{-1}) with concentrations (in mol L^{-1}), because in very dilute solutions of the solute $c_A V_A + c_B V_B = 1$, Eq. (5.12) can be transformed into

$$(\delta - \delta_{SA})/c_B = (\kappa_{SB}/\kappa_{SA})V_A(\delta_{SB} - \delta_{SA})$$
$$- [(\kappa_{SB}/\kappa_{SA})V_A - V_B](\delta - \delta_{SA}) \tag{5.13}$$

from which the ratio of the affinities, κ_{SB}/κ_{SA}, can be obtained. If light absorption in the UV is used as the property X and if only the complexed solute (subscripted c) but not the solvated solute (subscripted s) absorbs, then, from Eq. (5.10):

$$\chi_{SBc} = c_S\kappa_{SBc}c_B/[c_A\kappa_{SA} + (\kappa_{SBc} + \kappa_{SBs})c_B] \tag{5.14}$$

where $\kappa_{SBc} + \kappa_{SBs} = \kappa_{SB}$ is the total affinity of B to S. The absorbance at a given wavelength is $A = \varepsilon_{SBc}\chi_{SBc}$, hence, in analogy with Eq. (5.13):

$$A/c_B = (\kappa_{SBc}/\kappa_{SA})V_A\varepsilon_{SBc}c_S - [(\kappa_{SB}/\kappa_{SA})V_A - V_B]A \tag{5.15}$$

Equations (5.13) and (5.15) are written in the form of the Scatchard equation, but algebraic manipulation permits them to be transformed into the forms of the familiar Benesi–Hildebrand and Scott equations [26].

The COPS concept has been used by Nagy et al. [26] to interpret the ^{1}H NMR chemical shifts of the methyl protons of N-substituted phthalimides (S) in aromatic solvents (B) mixed with dichloromethane (A), showing Scatchard [Eq. (5.13)] as well as Benesi–Hildebrand plots. The equilibrium constant K for the adduct formation $S + B \rightleftharpoons SB$ in L mol^{-1} is obtained from

$$K = (\kappa_{SB}/\kappa_{SA})V_A - V_B \qquad (5.16)$$

Certain anomalies (such as apparently negative equilibrium constants) encountered in the use of the Benesi–Hildebrand, Scott, or Scatchard equations, depending on the relative values of K and V_B, can be explained by the COPS concepts. The distribution of the solute between the solvating (eventually, complexing) solvents A and B is given by:

$$Q = c_{SB}/c_{SA} = (\kappa_{SB}/\kappa_{SA})(V_B/V_A) \qquad (5.17)$$

neglecting volume changes of mixing and dissolution of the solute. This expression describes the partitioning in the homogeneous system [27,28]. The local mole fraction of the solute bound to B, called the "saturation factor" t, [26], is

$$x_{SB}^L = t = c_{SB}/(c_{SA} + c_{SB}) = Q/(1 + Q)$$
$$= (\kappa_{SB}/\kappa_{SA})/[1 + (\kappa_{SB}/\kappa_{SA})(V_B/V_A)] \qquad (5.18)$$

Curves of $x_{SB}^L = f(\phi_B)$ are shown in Fig. 5.1, for N-alkylphthalimides in aromatic solvents, obtained from ^1H NMR chemical shift data. The more bulky tetrachloro-N-methylphthalimide strongly prefers the 1-methyl-naphthalene solvent over dichloromethane, whereas the preference of N-methylphthalimide for toluene is much smaller. The COPS approach has also been applied to UV absorbance, kinetic, liquid–liquid distribution, and relaxation measurements [29,30].

   Acree et al. [31] have shown that the COPS expressions are similar to those derived from their NIBS approach (see later) to the solubility of nonelectrolytes in binary solvent mixtures. The mole fraction of the solute in the (dilute) saturated solution in the solvent mixture is given by

$$\ln x_S^{sat} = g_A \ln x_{S(A)}^{sat} + g_B \ln x_{S(B)}^{sat} \qquad (5.19)$$

where $g_A = 1 - g_B = c_A D_{SA}/(c_A D_{SA} + c_B D_{SB})$ [compare Eqs. (5.10) and (5.11)]. According to the NIBS approach, if the solvent–solvent interaction term can be neglected, the analogous expression is

$$\ln x_S^{sat} = f_A x_{S(A)}^{sat} + f_B x_{S(B)}^{sat} \qquad (5.20)$$

where $f_A = 1 - f_B = x_A/(x_A + x_B V_B/V_A)$ is a relative-volume-weighted mole fraction expression.

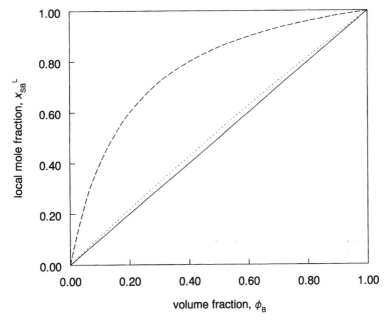

FIGURE 5.1 Preferential solvation of phthalimides in aromatic solvents from NMR data, according to the COPS theory: (- - - -) tetrachloro-N-methylphthalimide in 1-methylnaphthalene (B) + dichloromethane (A) mixtures; ($\cdots\cdots$) N-methylphthalimide in toluene (B) + dichloromethane (A) mixtures, at 298.15 K. (From Ref. 26.)

5.2.2 Thermodynamic Studies

The chemical potential or the partial molar enthalpy (from the enthalpy of solution) of a solute is, of course, sensitive to its environment and can be a good probe for its preferential solvation in a solvent mixture. Several approaches have been proposed for utilizing this fact. For instance, the NRTL expression used to describe the excess Gibbs energy in binary solvent mixtures (Section 2.C) was also used [11] for characterizing the solvation of a solute in such mixtures. The resulting equations are an extension of the NRTL approach to multicomponent systems:

$$x_{AS}^{L} = [x_A \exp(-\alpha(\tau_{AS} - \tau_{BS}))]/[x_B + x_A \exp(-\alpha(\tau_{AS} - \tau_{BS}))]$$

$$(5.21)$$

and $x_{BS}^L = 1 - x_{AS}^L$, where $\alpha = 0.30$ represents the nonrandomness factor (generally used for binary polar solvents) and the τ's are the interaction factors $\tau_{AS} = (g_{AS} - g_{AA})/RT$ and $\tau_{BS} = (g_{BS} - g_{BB})/RT$, the g's being the pairwise interaction (Gibbs) energies. This approach was applied to the preferential solvation of phenol blue in solvent mixtures [11] already studied spectroscopically (see earlier) and the pairwise interaction energies for the solute–solvent interactions were obtained from the solubilities of phenol blue in the neat solvents. Good agreement between the calculated NRTL curves and the spectroscopic results was obtained.

If the components of the binary mixed solvents have only a small excess Gibbs energy of mixing, the system can be described in terms of the *nearly ideal binary solvent* (NIBS) concept [32–34]. This assumes that only nonspecific interactions take place in the system. For such systems the expression in terms of volume fractions of the components (including the solute) is successful:

$$\Delta_{soln} G^\circ (\text{S in A} + \text{B}) = (1 - \phi_S^{sat})^2 [\phi_A \Delta_{soln} G^\circ (\text{S in A})$$
$$+ \phi_B \Delta_{soln} G^\circ (\text{S in B})]$$
$$- V_S (\phi_A V_A + \phi_B V_B)^{-1} G_{AB}^E \qquad (5.22)$$

where $\Delta_{soln} G^\circ = -RT \ln \phi_S^{sat}$, and ϕ_S^{sat} is the volume fraction of the solute in the saturated solution. When the NIBS approach is applied in cases in which the solubility is very small, $(1 - \phi_S^{sat})^2 \approx 1$, $\phi_A V_A + \phi_B V_B \cong V$, the molar volume of the mixed solvent and the coefficient of G_{AB}^E in Eq. (5.22) becomes V_S/V. The NIBS approach involves the standard molar Gibbs energies of solution and the excess molar Gibbs energy of mixing of the solvents and has no adjustable parameters. It could be applied to solubilities of naphthalene, iodine, tin(IV) iodide, p-benzoquinone, or benzil in mixtures of aromatic hydrocarbons or tetrachloromethane with aliphatic or alicyclic hydrocarbons with widely varying molar volumes, with a volume fraction of the solute, ϕ_S^{sat}, up to 0.35. Equation (5.22) has also been shown to be applicable to gas solubilities or, equivalently, to gas–liquid chromatographic partition coefficients of the gaseous solute in a mixed solvent stationary phase [33]. In all cases where the NIBS approach is applicable, there is, naturally, no significant preferential solvation of the solute by any of the solvent components.

Acree and coworkers [35] studied the solubility of large molecules, such as anthracene, in binary solvent mixtures involving alcohols, where one of the components or both is capable of self-association. They considered both Huyskens' mobile order theory [36] and the Kretschmer–Wiebe self-association model [37] for this purpose. The mobile order theory considers molecules moving continuously in mobile domains defined by the volume available for them, being the total volume of the liquid divided by the number of molecules of the kind considered. They do so unless restrained from it by hydrogen bonding to some other molecules. The fraction of time, γ_A, that the alcohol molecule A is not involved in hydrogen bonding is the key variable of the method. It is related to the equilibrium constant for the hydrogen bonding, K_A, by $\gamma_A = 1/(1 + K_A \phi_A / V_A)$, where ϕ_A is the volume fraction of the alcohol and V_A is its molar volume. The self-association model assumes the solvent molecules forming aggregates, with a single equilibrium constant describing the addition of a new molecule to an already existing aggregate. These aggregates, in the case of alcohols, are hydrogen-bonded polymeric chains, and the equilibrium constant is taken to be on the molar scale. When the mixture contains two alcohols, both homogeneous and heterogeneous polymeric chains are considered, but it is expedient to use a single equilibrium constant to describe all kinds of aggregates ($5 \ \mathrm{dm^3 \ mol^{-1}}$ in the cases studied). Rather complicated expressions resulted from these treatments, but both were capable of expressing the solubility in the mixtures well, given the solubilities of the solute in the neat solvents.

The NIBS approach also provides an expression for the limiting (infinite dilution) partial derivative of the chemical potential of the solute with respect to the number of moles of a solvent component:

$$\lim(c_S \to 0)(\partial \mu_S / \partial n_A)_{T,P,n_B} = -RT(V_S V_A / V^2)(1 - V_A / V_B)$$
$$+ (V_S V_A V_B / V^2)$$
$$\times [A_{SA} - A_{SB} + (\phi_A - \phi_B)A_{AB}]$$

$$(5.23a)$$

The NIBS expression for the solubility of the solute S is

$$RT[\ln(a_S^{id}/\phi_S^{sat}) - (1 - \phi_S^{sat})(1 - V_S/V)]$$
$$= (1 - \phi_S^{sat})^2 V_S \times [\phi_A A_{SA} + \phi_B A_{SB} - \phi_A \phi_B A_{AB}] \quad (5.23b)$$

where a_S^{id} is the ideal solubility, i.e., the activity of the solid in equilibrium with the saturated solution, obtained from the enthalpy of fusion and the heat capacity difference between the solid and the (undercooled) liquid solute. The volume fractions ϕ pertain to the solute in the saturated solution, ϕ_S^{sat}, to the components of the mixture, and the total volume of the solution is $V = x_A V_A + x_B V_B$. The A_{SA} and A_{SB} parameters are obtained from Eq. (5.23b) setting ϕ_B and ϕ_A, respectively, to zero, i.e., the solubilities in the individual neat solvents, and A_{AB} is then obtained from the solubility in the mixture. It is important to note that Eq. (5.23a) bears a relationship to the expression for $\lim(c_S \to 0)(\partial \mu_S / \partial n_A)_{T,P,n_B}$, Eq. (5.29), obtained from the Kirkwood–Buff integrals, that describes the preferential solvation of the solute in the binary solvent mixture [38].

The local mole fraction of the solvent component A in the surroundings of the solute S, x_{AS}^L, expressing the preferential solvation of S in the mixture, can be obtained from the Kirkwood–Buff space integral of the pair correlation function, $g_{AS}(r)$, employing the *inverse Kirkwood–Buff integral* (IKBI) method. The function $g_{AS}(r)$ expresses the probability of finding the center of a molecule of solvent A in an infinitesimal volume element at a distance r from the center of a particle of solute S, located at a certain position in the mixture, averaged over all the possible mutual orientations. The (Kirkwood–Buff) integral is

$$G_{AS} = \int_0^\infty [g_{AS}(r) - 1]4\pi r^2 \, dr \tag{5.24}$$

and expresses the *affinity* of solvent A for the solute S. The function $g_{AS}(r)$ differs from unity only in a region a few molecular diameters away from the center of S, so that the main contribution to G_{AS} comes from this *correlation region*, extending to the distance R_{cor}. The volume of this region, where the solute S exerts an influence on its surroundings, is $V_{cor} = (4\pi/3)R_{cor}^3$. The average number of A molecules in the correlation region around S, N_{AS}, is given by the product of the bulk number density of A, ρ_A, and the integral of the pair correlation function of A and S in the correlation volume, according to Ben-Naim [39]:

$$\begin{aligned}
N_{AS} &= \rho_A \int_0^{R_{cor}} g_{AS}(r)4\pi r^2 \, dr \\
&= \rho_A \int_0^{R_{cor}} [g_{AS}(r) - 1]4\pi r^2 \, dr + \rho_A \int_0^{R_{cor}} 4\pi r^2 \, dr \\
&\cong \rho_A \int_0^\infty [g_{AS}(r) - 1]4\pi r^2 \, dr + \rho_A (4\pi/3)R_{cor}^3 \\
&= \rho_A G_{AS} + \rho_A V_{cor} \tag{5.25}
\end{aligned}$$

The near equality in the third line of Eq. (5.25) arises from substitution of infinity for R_{cor} in the upper limit of the first integral, noting that only an entirely negligible contribution to this integral arises from the region beyond R_{cor}, where $g_{AS}(r) \approx 1$. The average number of B molecules in this region is similarly $N_{BS} = \rho_B G_{BS} + \rho_B V_{cor}$. The local mole fraction of component A around S, understanding that "local" pertains to the correlation region, is

$$x^L_{AS} = N_{AS}/(N_{AS} + N_{BS})$$

$$= [\rho_A G_{AS} + \rho_A V_{cor}]/[\rho_A G_{AS} + \rho_B G_{BS} + (\rho_A + \rho_B) V_{cor}]$$

$$= [x_A G_{AS} + x_A V_{cor}]/[x_A G_{AS} + x_B G_{BS} + V_{cor}] \qquad (5.26)$$

where division by $(\rho_A + \rho_B)$ and substitution of $x_A = \rho_A/(\rho_A + \rho_B)$ are used in the last line. The preferential solvation parameter is now obtained from Eq. (5.26) as

$$\delta x_{AS} = x_A x_B (G_{AS} - G_{BS})/[x_A G_{AS} + x_B G_{BS} + V_{cor}] \qquad (5.27)$$

The numerator of Eq. (5.26) may be taken as the limiting preferential solvation, when δx_{AS} is expanded as a power series in V^{-1}_{cor} and the limit at $V^{-1}_{cor} = 0$ is taken:

$$\delta x^{\circ}_{AS} = x_A x_B (G_{AS} - G_{BS}) \qquad (5.28)$$

This shows whether S is surrounded preferentially by molecules of A (if $G_{AS} > G_{BS}$) or by molecules of B because the denominator in Eq. (5.27) is always positive. Thus, the relative magnitudes of the Kirkwood–Buff integrals G_{AS} and G_{BS}, which are measures of the affinities of A and B to S, determines the preferential solvation of the solute in the binary solvent mixture of A and B.

It remains to indicate how these Kirkwood–Buff integrals are obtained from experimental data. The limiting partial derivative of the chemical potential of S with respect to the number of moles of solvent component B, n_B, is

$$\lim(x_S \to 0)(\partial \mu_S/\partial n_B)_{T,P,n_A} = RT(n/V)^2 (G_{BS} - G_{AS})/\eta \qquad (5.29)$$

where n is the number of moles of mixed solvent and $\eta = (n/V) + (n_A n_B) V^{-2} [G_{AA} + G_{BB} - 2 G_{AB}]$. This expression can be modified as follows [39], using the standard molar solvation Gibbs energy of the solute S, present at infinite dilution in the solvent mixture, $\Delta_{solv} G^{\circ}_S$. The derivative of $\Delta_{solv} G^{\circ}_S$ with respect to the composition gives the desired difference $G_{AS} - G_{BS}$:

$$\lim(x_S \to 0)[\partial \Delta_{solv} G^{\circ}_S/\partial x_B]_{T,P} = (RTQ/V)(G_{AS} - G_{BS}) \qquad (5.30)$$

The quantity Q depends on the properties of the binary solvent mixture in the absence of the solute as for the binary solvent mixtures, Eq. (4.26) in section 4.B:

$$Q = x_j[\partial(\mu_i/RT)/\partial x_i]_{T,P} = x_j[\partial \ln a_i/\partial x_i]_{T,P} = 1 + x_j[\partial \ln f_i/\partial x_j]_{T,P}$$
$$= 1 + x_A x_B[\partial^2(G^E/RT)/\partial x_A^2]_{T,P} \tag{5.31}$$

In the first row of Eq. (5.31) the component i can be selected as either A or B, according to the availability of the data. Rather than use the solvation Gibbs energy in Eq. (5.30), in cases where the solute S is nonvolatile, the standard molar Gibbs energy of transfer of S, $\Delta_{tr}G_S^\circ(S)$, from a reference solvent to the binary mixture may be employed, where $\Delta_{tr}G^\circ{}_S(S)$ stands for $\lim(x_S \to 0)[\Delta_{tr}G_S(S, A \to A + B)]$. The reference solvent may be either another solvent altogether or simply neat A, yielding the difference of the Kirkwood–Buff integrals:

$$G_{AS} - G_{BS} = [\partial \Delta_{tr}G_S^\circ(S)/\partial x_B]_{T,P}/(RTQ/V) \tag{5.32}$$

Else, transfer from neat B may be used: $\Delta_{tr}G_S^\circ(S, B \to A + B)$ and again its derivative with respect to the mole fraction of solvent B. The individual Kirkwood–Buff integrals may be obtained from the standard partial molar volume of the solute, V_S°, and the partial molar volume of the solvents in the mixture in the absence of the solute, $V_A V_B$:

$$G_{AS} = RT\kappa_T - V_S^\circ + x_B V_B[\partial \Delta_{tr}G_S^\circ/\partial x_B]/Q \tag{5.33a}$$
$$G_{BS} = RT\kappa_T - V_S^\circ + x_A V_A[\partial \Delta_{tr}G_S^\circ/\partial x_B]/Q \tag{5.33b}$$

where κ_T is the isothermal compressibility of the solvent mixture in the absence of the solute. The preferential solvation parameter of the solute S in dilute solution in the mixture of solvents A and B can, thus, be obtained from Eqs. (5.28) and (5.32) [39]. As seen, this requires the derivative of its transfer Gibbs energy into the mixture with respect to the solvent composition, as well as information on the mixing of the solvents in the absence of the solute, i.e., the quantity Q, Eq. (5.31), which also involves derivative functions. The necessity of using derivatives of experimental quantities detracts from the accuracy of the resulting preferential solvation parameter.

Instead of obtaining the Kirkwood–Buff integrals from thermodynamic data, Bagchi et al. [40] showed that Eq. (5.27) can be inverted to yield these integrals by the use of δx_{AS} values, obtained from spectroscopic measurements according to Eqs. (5.3) and (5.4). However, an uncertainty in this procedure is caused by the not necessarily justified assumption of the linear dependence of the measured quantities on the composition of the solvation shell involved in Eq. (5.3). Certain further assumptions and simplifications cause additional uncertainties in this approach. Hence, the derived Kirkwood–Buff integrals (rather their ratios to the correlation volume V_C) are not very reliable.

An alternative method for obtaining local mole fractions or preferential solvation parameters for a solute in a binary solvent mixture, not involving derivative functions, may be desirable. This can be found in the quasi-lattice quasi-chemical (QLQC) method proposed by Marcus [41], which depends on a model and hence is less rigorous than the IKBI method already described. The quasi-lattice part of the model assumes that the particles of the solute S and of the solvents A and B are distributed on sites of a quasi-lattice of the solution that is characterized by a lattice parameter Z. This parameter specifies the number of neighbors each particle has and is independent of the nature of these particles. The configurational energy of the system is determined solely by the sum of the pairwise interaction energies of nearest neighbors: e_{AS}, e_{BS}, e_{SS}, e_{AA}, e_{BB}, and e_{AB} weighted according to the numbers of neighbors of each kind found in the solution. These energies are assumed to be independent of the natures of any other neighbors the partners of a neighboring pair may have. The mixing of the particles on the quasi-lattice is assumed to take place without a volume change of the system, i.e., $V^E = 0$, and an ideal entropy of mixing is assumed, i.e., $S^E = 0$. Hence, the Gibbs, Helmholtz, and configurational energies are equal to each other. In the ternary mixture there are N_{AS} nearest neighbor pairs of particles of the solute S and molecules of the solvent A and N_{BS} nearest neighboring pairs of S and B, as well as N_{SS} pairs of neighboring S particles. The quasi-chemical assumption of the model specifies the number of nearest neighbor pairs of molecules in the mixture, according to Guggenheim [42],

$$N_{AS}^2 = 4N_{AA}N_{SS}\exp[\Delta e_{AS}/k_B T] \tag{5.34}$$

where $\Delta e_{AS} = e_{AA} + e_{SS} - 2e_{AS}$ and k_B is Boltzmann's constant. A similar expression can be written for N_{BS}^2, on changing subscript A to B. The local mole fraction of A around S can obviously be written in terms of the numbers of nearest neighbors as [41]

$$x_A^L = N_{AS}/(N_{AS} + N_{BS}) = 1/[1 + N_{BS}/N_{AS}] \tag{5.35}$$

This can be rewritten using the square roots of the expressions on each side of Eq. (5.34) and the corresponding expressions for N_{BS}^2 as

$$x_A^L = 1/[1 + (N_{BB}/N_{AA})^{1/2} \exp(\Delta e_{ABS}/2k_B T)] \tag{5.36}$$

where $\Delta e_{ABS} = \Delta e_{BS} - \Delta e_{AS}$. Local mole fractions and preferential solvation parameters can thus be obtained from Eq. (5.36), provided that the ratio N_{BB}/N_{AA} the solvation energy difference Δe_{ABS} are available.

For a very dilute solution of S in the solvent mixture, the ratio N_{BB}/N_{AA} depends only on the properties of the binary solvent mixture, containing N_A molecules of A and N_B molecules of B. Let the sum of nearest neighbors in the mixture, disregarding the solute, be

$$L = N_{AA} + N_{BB} + N_{AB} = (Z/2)(N_A + N_B) \tag{5.37}$$

Then:

$$ZN_A = 2N_{AA} + N_{AB} \tag{5.38}$$

$$N_{AA}/L = x_A - N_{AB}/2L \tag{5.39}$$

and similarly for B. Therefore:

$$N_{BB}/N_{AA} = (x_B - N_{AB}/2L)/(x_A - N_{AB}/2L) \tag{5.40}$$

and it remains to express $N_{AB}/2L$ in terms of measurable quantities. This is done as discussed previously (Section 4.2):

$$N_{AB}/2L = [1 - \{1 - 4x_A x_B (1 - \exp(-\Delta e_{AB}/k_B T))\}^{1/2}]/$$
$$[2(1 - \exp(-\Delta e_{AB}/k_B T))] \tag{5.41}$$

where the binary interaction energy difference Δe_{AB} is obtained from the excess Gibbs energy of mixing at the equimolar composition, $G_{AB(x=0.5)}^E$ [41]:

$$\exp(\Delta e_{AB}/k_B T) = \{[2 \exp(-2G_{AB(x=0.5)}^E/ZRT)] - 1\}^2 \tag{5.42}$$

For the calculation of the local mole fractions of the solvents around the solute S it is still necessary to specify the interaction energy difference Δe_{ABS}. This is equal to the difference between the standard molar Gibbs energies of solvation of S divided by Z times Avogadro's number, N_{Av}:

$$\Delta e_{ABS} = [\Delta_{solv} G^\circ(S, A) - \Delta_{solv} G^\circ(S, B)]/ZN_{Av} \tag{5.43}$$

As for the IKBI method discussed earlier, standard molar Gibbs energies of transfer of S from a reference solvent (which may be one of the neat components of the mixed solvent) into the mixture may be substituted for the Gibbs energies of solvation. Equations (5.40) to (5.43) provide with Eq. (5.36) all that is necessary for the calculation of the local mole fractions and the preferential solvation parameters. For the QLQC model, thus, the experimental value of the molar excess Gibbs energy of mixing of the equimolar solvent mixture, $G^E_{(x=0.5)}$; the lattice parameter Z, obtained as a fitting parameter of the $G^E(x)$ curve; and the Gibbs energy of transfer of the solute from the one neat solvent into the other neat solvent uniquely determine the behavior of the solute in the mixed solvent system, i.e., the local environments of its particles and its preferential solvation [41].

It is, furthermore, possible to use the QLQC method to calculate the standard molar Gibbs energy of transfer of the solute from a reference solvent into the binary solvent mixture. These calculated values can then be compared with the experimentally determined ones. The quasi-chemical part of the QLQC method, as developed by Hagemark [43], specifies the excess chemical potential of a component i (S, A, or B) in the solution as

$$\mu^E_i = (RTZ/2)\ln(N_{ii}/x^2_i L) \tag{5.44}$$

This can be written for the solute, using quasi-chemical expressions similar to Eq. (4.14), as

$$\mu^E_S = -(ZRT)\ln[(N_{AA}/L)^{1/2}\exp(\Delta e_{AS}/2k_B T)$$
$$+ (N_{BB}/L)^{1/2}\exp(\Delta e_{BS}/2k_B T)] \tag{5.45}$$

Equation (5.44) can be rewritten in terms of separate solvent–solvent and solute–solvent interactions at the solvent composition x (either x_A or x_B) [41]:

$$\mu^E_S(x) = -G^E(x) + x_A\mu^E_{S(A)} + x_B\mu^E_{S(B)} - f(x) \tag{5.46}$$

Here $G^E(x)$ is, as before, the excess Gibbs energy of mixing the solvents in the absence of the solute, $\mu^E_{S(A)}$ is the excess chemical potential of the solute S in neat solvent A, and similarly, for $\mu^E_{S(B)}$, with respect to neat B. The function $f(x)$ is given by

$$f(x) = ZRT[x_B\ln\{x_B + x_A y\exp(\Delta)\} + x_A\ln\{x_A + x_B y^{-1}\exp(-\Delta)\} \tag{5.47}$$

where

$$y = (x_A/x_B)(N_{BB}/N_{AA})^{1/2} \tag{5.48}$$

calculated from Eqs. (5.40) to (5.42) already given earlier,

$$\Delta = (\Delta e_{SA} - \Delta e_{SB})/k_B T$$

$$= [\Delta_{tr} G_S^\circ(S, R \to A) - \Delta_{tr} G_S^\circ(S, R \to B)]/ZRT \qquad (5.49)$$

The standard molar Gibbs energy of transfer of the solute from the reference solvent R into the mixed solvent A + B is now obtained from Eq. (5.46) by

$$\Delta_{tr} G_S^\circ(S, R \to A + B) = x_A \Delta_{tr} G_S^\circ(S, R \to A) + x_B \Delta_{tr} G_S^\circ(S, R \to B)$$

$$- G^E(x) - x_A x_B f(x) \qquad (5.50)$$

In Eqs. (5.49) and (5.50) R is a reference solvent, and if it is either neat solvent A or B, then the corresponding $\Delta_{tr} G_S^\circ$ is, of course, zero. The standard molar Gibbs energies of transfer should be given on the mole-fraction concentration scale for the transition from Eq. (5.46) to Eq. (5.50) to hold. Thus, $\Delta_{tr} G_S^\circ(S, R \to A + B)$ can be calculated from the standard molar Gibbs energies of transfer of the solute from the reference solvent R into the two neat solvents and the excess Gibbs energy of mixing of the two solvents in the absence of the solute. It also requires the function $f(x)$, which depends, in turn, on the standard molar Gibbs energies of transfer of the solute from the reference solvent R into the two neat solvents and the quasi-lattice parameter, Z, via the quantity Δ, and the function y [41]. The quasi-lattice parameter Z remains as the only fitting parameter.

As mentioned previously, the model specifies a constant value of Z, irrespective of the occupancy of the quasi-lattice sites by particles of S, A, or B. Hence, if it was determined by the QLQC model for the mixtures of the two solvents A and B [Section 4.B, Eqs. (4.19) to (4.22)], it should hold for the transfer of any solute S into these mixtures. As this is not necessarily the case, this feature is a weakness of the QLQC approach. A further weakness is the assumption of a zero excess entropy of mixing, that is, random mixing of the particles of S, A, and B on the quasi-lattice sites, because preferential solvation produces a negative entropy of mixing. Furthermore, in cases in which the solute solubility is such that the standard molar Gibbs energy of transfer of the solute from the one neat component to the other is small, near $1\,kJ\,mol^{-1}$, or if the $\Delta_{tr} G^\circ(S, A \to A + B) = f(x)$ curve has an extremum, the QLQC method appears not to yield correct values of $\Delta_{tr} G^\circ$.

For systems involving solid or gaseous solutes in a binary solvent mixture, the Gibbs energy of solvation or of transfer of the solute is required, as shown in Eqs. (5.30), (5.32), (5.33), (5.43), (5.49), and (5.50). The applications of the IKBI and QLQC methods have often been to the solubilities of sparingly soluble solutes in binary solvent mixtures. The transfer of the solute is, thus, not from a reference solvent R but from the solid or gaseous state. In the case of solids, the nonformation of crystalline solvates has to be postulated so that the phase in equilibrium with the saturated solution is the same throughout the solvent composition range. It turns out that the QLQC method is applicable to cases where there are appreciable differences between the solubilities in the two neat solvents, so that $\Delta_{tr}G^{\circ}(S, A \rightarrow B)$ is at least a few kJ mol^{-1} and the $\Delta_{tr}G^{\circ}(S, A \rightarrow A + B) = f(x)$ curve is monotonous. Two such examples are shown in Figs. 5.1 and 5.2, where the continuous curves have been calculated by the QLQC method and the discrete symbols are the experimental data. Figure 5.2 pertains to the solubility of magneson [4-(4-nitrophenylazo)-1,3-benzenediol] in mixtures of water and N-methylformamide [44],

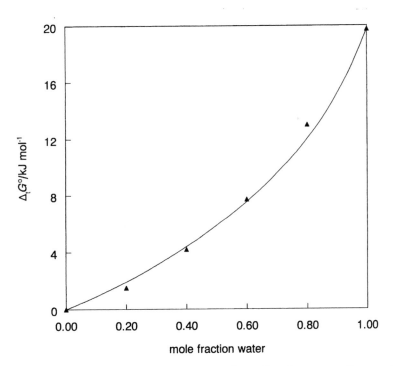

FIGURE 5.2 Standard molar Gibbs energy of transfer of magneson from water to aqueous N-methylformamide. (▲) Obtained from the Solubility; (———) calculated by the QLQC method with $Z = 4$. (From Ref. 44.)

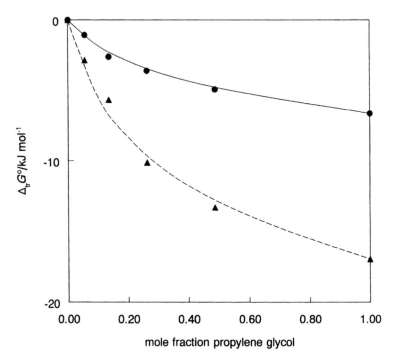

FIGURE 5.3 Standard molar Gibbs energies of transfer of ethyl-*p*-aminobenzoate (•) and dodecyl-*p*-aminobenzoate (▲) from water to aqueous 1,2-propanediol. The curves are calculated by the QLQC method with $Z = 3$ (———) and $Z = 2$ (— — —). (From Ref. 45.)

and Fig. 5.3 pertains to the solubilities of ethyl or dodecyl *p*-amino-benzoate in mixtures of water and 1,2-propylene glycol [45]. The local mole fractions of the organic component of the mixed solvent around the solute, $x_{\text{org}}^{\text{L}}$, have been calculated, and in all these cases the solute prefers this component ($x_{\text{org}}^{\text{L}} > x_{\text{org}}$) over water. The actual numbers of the mole-cules of the two components in the near environment of the solute are given by Zx_{A}^{L} and Zx_{B}^{L}.

The solubilities of solid organic compounds (of pharmaceutical significance) in aqueous (A) mixtures with organic solvents (B) were modeled by Khossravi and Connors [46]. The model includes the work required for the creation of the cavity in the solvent mixture to accommodate the solute and the stepwise replacement of the water in the solvation shell, as in Eq. (5.6). The cavity term depends on the composition of the solvation shell and utilizes the bulk surface tensions σ and a quantity depending on the surface area of the solute but appearing as a fitting parameter C. The solvent replacement term is taken as a two-step replacement ($N=2$) of one B molecule for one A molecule, with the equilibrium constants, see Eq. (5.6), $\beta_1 = K_{A/B1}$, $\beta_2 = K_{A/B1}K_{A/B2}$. The resulting expression for the standard molar Gibbs energy of transfer of the solute from water to the mixture is

$$\Delta_t G^\circ(A \rightarrow A + B) = RT[(\ln \beta_1 + C(\sigma_B - \sigma_A)/2)\beta_1 x_A x_B$$
$$+ (\ln \beta_2 + C(\sigma_B - \sigma_A))\beta_2 x_A x_B^2]/$$
$$[x_A^2 + \beta_1 x_A x_B + \beta_2 x_B^2] \qquad (5.51)$$

This expression, with three fitting parameters, C, β_1, and β_2, was applied to 33 solutes in eight aqueous cosolvent mixtures with good results (some of the systems required extension to $N=3$ with a fourth parameter, β_3).

The solubilities of gases in solvent mixtures often show non-monotonous and/or very asymmetric $\Delta_{tr} G^\circ(S, A \rightarrow A + B) = f(x)$ curves. Hence, they cannot be treated in terms of the QLQC method because the assumptions at the basis of its model break down. They can, however, be interpreted in terms of the IKBI method, at least qualitatively, because this method considers the slope of the $\Delta_{tr} G^\circ(S, A \rightarrow A + B) = f(x)$ curve. Consider, for example, the solubilities of argon and of carbon dioxide (S) in aqueous (A) ethanol (B). The Ostwald coefficients, γ_S, of the gas solubility are translated into standard molar Gibbs energies of solution, $\Delta_{soln} G^\circ = - RT \ln \gamma_S$, and these, in turn, into standard molar Gibbs energies of transfer from water into aqueous ethanol: $\Delta_{tr} G^\circ(S, A \rightarrow A + B) = \Delta_{soln} G^\circ(S, A + B) - \Delta_{soln} G^\circ(S, A)$ as a function of x_B [47]. The results have an appreciable temperature dependence, and $\Delta_{tr} G^\circ$ starts out positive as x_B increases, turning towards negative values at higher ethanol contents. The resulting limiting preferential solvation parameters (at $V_{cor}^{-1} \rightarrow 0$) $\delta x_{AS}^\circ = x_A x_B(G_{AS} - G_{BS})$ are shown for carbon dioxide as the solute and water as the solvent component A in Fig. 5.4. At low ethanol concentrations water is the preferred solvent around the carbon dioxide, as it also is at ethanol concentrations above 40 mole %, but at intermediate ethanol contents it is the organic solvent that is preferred near the solute. This effect is enhanced at the higher temperature and finds its explanation in the structure of aqueous ethanol mixtures, as similar

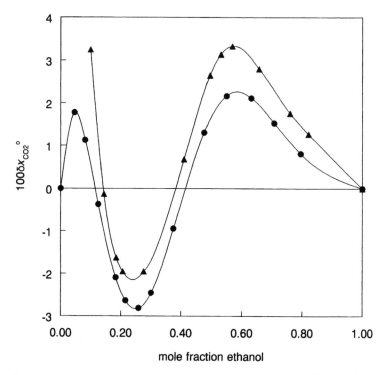

FIGURE 5.4 Preferential solvation parameters of carbon dioxide in aqueous ethanol obtained by the IKBI method: (●) at 288.15 K; (▲) at 308.15 K. (From Ref. 47.)

behavior was found for argon as the solute, which is devoid of specific interactions with the components of the mixture.

Because both the IKBI and QLQC methods utilize the Gibbs energy of transfer of the solute from a reference solvent to the solvent mixtures, primary data other than solubilities can be used, such as liquid–liquid distribution. A liquid must then be found that interacts sufficiently strongly with the solute but is immiscible with the solvent mixture under discussion for the solute to distribute between the two liquid phases. Such a solvent is *n*-octane, which can be used with aqueous mixtures of very polar solvents, such as methanol or dimethylsulfoxide, and, generally, not too hydrophilic solutes. Two substantially immiscible liquid phases are formed in such systems, between which solutes may distribute.

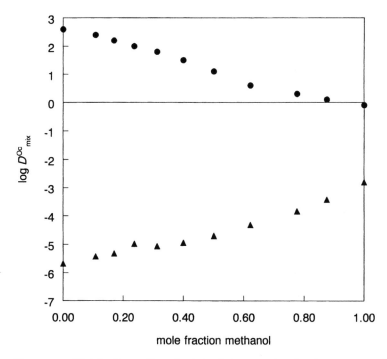

FIGURE 5.5 Distribution ratios of solutes between octane and aqueous methanol, calculated from group increments at 20°C: (•) iodobenzene; (▲) ethanolamine. (From Ref. 48.)

The distribution ratios between octane and aqueous methanol mixtures, $\log D_{mix}^{Oc}$, are illustrated in Fig. 5.5 for the very hydrophobic iodobenzene and the very hydrophilic ethanolamine as solutes. These were calculated from the group increments reported by Leshchev [48]. The former solute prefers the octane phase, but less and less so as the methanol content increases, whereas the latter solute strongly prefers the aqueous mixture, but again less and less so as the methanol content increases. The slopes of the curves can thus be interpreted, according to Eqs. (5.32) and (5.28) of the IKBI method, because $\Delta_t G^\circ(S, A \to A + B) = -RT(\log D_{mix}^{Oc} - \log D_A^{Oc})$. In terms of the preferential solvation of the solutes by the components of the mixed aqueous methanol solvent, iodobenzene is preferentially solvated by the methanol and ethanolamine by the water throughout the composition range.

There appear to be no widely used expressions for the relation of the enthalpies of solution of nonelectrolyte solutes in binary mixed solvents to their preferential solvation, although suitable data for this purpose have been published (e.g., [49,50]). Neither the IKBI nor the QLQC thermodynamic approach is readily amenable for such use, the latter because the assumptions at its basis break down in view of the rather non-symmetrical enthalpy curves (but see the following).

An approach to preferential solvation that is similar in some respects to the COPS method but differs in other respects is the extended coordination model (ECM). This was originally formulated for the coordination of ions in binary solvent mixtures (see later) [51,52] but was later extended to nonelectrolyte solutes by Cox and Waghorne [53] and deals with the enthalpy of transfer of the solute from a reference solvent to the binary solvent mixture. The symbols used by the authors are employed here, even when they differ from symbols for the same quantities used elsewhere in this chapter, so that the original publications can be readily followed. The model considers the first coordination/solva-solvation shell of the solute S with a total of n solvent molecules [lowercase n is used instead of the capital N used previously, Eqs. (5.6) and (5.7)]. The local mole fractions of the two solvents are $x_{AS}^L = n_{AS}^L/n$ and $x_{BS}^L = n_{BS}^L/n$ (and $n_{AS}^L + n_{BS}^L = n$) and the preferential solvation parameter is $p = (x_B^L/x_A^L)/(x_B/x_A)$ [$= K_{AB/S}$ as before, Eq. (5.5)]. The solvent–solvent interactions in the first solvation shell are disrupted to some extent ($0 < \alpha < 1$). Also, the ratio $x_{BS}^L/x_{AS}^L = n_{BS}^L/n_{AS}^L \approx N_B/N_A$, where in the last, approximate equality $N(\geq n) = N_{BS} + N_{AS}$ is the number of solvent molecules of both kinds beyond the first solvation shell, of which the solvent–solvent interactions are disrupted to some extent ($0 < \beta < 1$) due to the presence of the solute in their vicinity. The combined quantity ($\alpha n + \beta N$) is a parameter of this model. Further quantities that enter this model, obtainable from direct experimental data, are the difference in the enthalpies of interaction of the two solvents, equated with the difference in their standard enthalpies of vaporization, $\Delta H^* = (\Delta_v H_A^\circ - \Delta_v H_B^\circ)$, and the difference in the enthalpies of interaction of the solute with the two solvents, equated with the difference in their standard enthalpies of solvation $\Delta_t H^\circ(S, A \to B) = (\Delta_{solv} H_B^\circ - \Delta_{solv} H_A^\circ) = (\Delta_{sln} H_B^\circ - \Delta_{sln} H_A^\circ)$. Here, $\Delta_t H^\circ$, the standard molar enthalpy of transfer, can be taken as the difference between the standard molar enthalpies of dissolution, $\Delta_{sln} H^\circ$. Further quantities that are needed are the relative partial molar enthalpies of the two solvents in the mixture, l_A and l_B (see Section 2.3). The model then specifies that the enthalpy of transfer of the solute into the binary mixed solvent $\Delta_t H^\circ(S, R \to A + B)$ is

$$
\begin{aligned}
\Delta_t H^\circ(S, R \to A + B) = {} & p x_B (x_A + p x_B)^{-1} [\Delta_t H^\circ(S, A \to B) \\
& + (\alpha n + \beta N)\Delta H^*] + (\alpha n + \beta N)(x_A + p x_B)^{-1} \\
& \times [x_A l_A + p x_B l_B]
\end{aligned} \tag{5.52}
$$

The unknown parameters, p and $\alpha n + \beta N$, must be obtained from the experimentally determined $\Delta_t H^\circ(S, R \to A + B)$ values, together with experimental data for ΔH^*, $\Delta_t H^\circ(S, A \to B)$, l_A, and l_B, the latter two quantities as functions of the composition. Because of the nonlinear dependence on p, the accuracy of the fitting cannot be high. However, in some cases the parameter p could be compared with the quantity independently determined for the preferential solvation of the functional group of the solute by infrared spectroscopy with good agreement. In the latter studies, the stepwise replacement equilibria, $SA_2 + B \rightleftharpoons SAB + A$ (equilibrium constant K_1) and $SAB + B \rightleftharpoons SB_2 + A$ (equilibrium constant K_2), were considered and their equilibrium constants were given for certain amides in binary solvent mixtures by Symons and coworkers [21–24]; see earlier. The preferential solvation parameter p can be obtained from these equilibrium constants as

$$
p = [K_1 + 2K_1 K_2 (x_B/x_A)]/[1 - K_1 K_2 (x_B/x_A)^2] \tag{5.53}
$$

In some cases the partial replacement of A by B can be ignored and Eq. (5.53) can be simplified by elimination of K_1 in the numerator, the total replacement equilibrium constant being $\beta_2 = K_1 K_2$, leading to $p = 2\beta_2(x_B/x_A)/[1 - \beta_2(x_B/x_A)^2]$.

Values of p and $\alpha n + \beta N$ for a number of solutes solvated in binary solvent mixtures are shown in Table 5.2. In cases where $p < 1$, it is solvent B that is preferred by the solute over solvent A, and vice versa for $p > 1$. When $p \approx 1$, neither component of the mixture is preferred, and the near environment of the solute molecule has the same composition as the bulk solvent. Amides and dimethylsulfoxide generally prefer the protic, hydrogen-bonding, solvent around them, except in some water-rich mixtures. Other solutes may prefer one or the other component of the mixture, according to their polarity and hydrogen bond accepting properties.

The stepwise solvent replacement approach has been applied [54] to the enthalpies of solution of polyols in aqueous (W) dimethylformamide (DMF) mixtures. De Visser and Somsen [55] showed that the

TABLE 5.2   Parameters for the Preferential Solvation of Nonelectrolytes in Binary Solvent Mixtures

Solvent mixture	Solute	p	$(\alpha n + \beta N)$
Water (A) +	Formamide	1.8	1.3
dimethylsulfoxide (B)[a]	N-Methylformamide	0.85[b]	
	N-Methylacetamide	0.70[b]	
	N,N-Dimethylformamide	0.96	3.1
	N,N-Dimethylacetamide	0.8	3.8
	Acetic acid	2.9	1.9
	Methyl acetate	1.6	3.3
	Ethyl acetate	1.5	4.1
	t-Butyl acetate	1.7	4.8
	Acetone	1.1	3.1
	Acetaldehyde diethylacetal	1.1	6.9
	Ethyl vinyl ether	2.3	3.7
	Benzene	2.8	2.2
	Trimethyl phosphate	1.8	3.1
	Acetonitrile	1.1	1.9
Acetonitrile (A) +	Dimethylsulfoxide	0.4	4.0
methanol (B)[c]	Dimethylsulfone	1.8	2.0
	Dimethyl sulfate	1.7	0.8
	Acetic acid	0.6	3.3
	Methyl acetate	1.0	0.9
	Formamide	0.6	2.7
	N-Methylformamide	0.5	2.5
	N,N-Dimethylformamide	0.6	2.8
	N,N-Dimethylacetamide	0.6	4.1
	Propylene carbonate	1.4	0.8
	Water[d]	0.56	1.4
	n-Propanol[d]	0.50	0.8
	t-Butanol[d]	0.67	0.7
	n-Octanol[d]	0.50	0.58
Acetonitrile (A) +	Formamide	1.3,0.3	2.9,1.6
water (B)[e]	N-Methylformamide	0.5,0.4	9.8,1.6
	N,N-Dimethylformamide	0.4,0.6	16.3,1.6
Methanol (A) +	Formamide	3.2,0.2	2.1,0.4
water (B)[f]	N-Methylformamide	1.1,0.2	6.4,1.1
	N,N-Dimethylformamide	1.0,0.2	8.0,1.9
	N-Methylpyrrolidin-2-one	1.3,0.2	9.5,2.8
Ethanol (A) +	Formamide	1.7,0.2	3.8,3.8
water (B)[f]	N-methylformamide	1.3,0.3	6.6,4.4
	N,N-Dimethylformamide	1.1,0.3	9.5,1.6
	N-Methylpyrrolidin-2-one	1.6,0.3	12.8,0.8

TABLE 5.2   Continued

Solvent mixture	Solute	p	$(\alpha n + \beta N)$
n-Propanol (A) +	Formamide	1.0,0.5	4.5,5.9
water (B)[f]	N-Methylformamide	0.6,0.3	7.8,2.3
	N,N-Dimethylformamide	0.6,0.3	12.1,0.7
	N-Methylpyrrolidin-2-one	0.6,0.2	17.8,0.0
	Urea[g]	1.0	4.7
t-Butanol (A) +	Formamide	1.7	3.9
water (B)[f,g]	N-Methylformamide	1.1	6.3
	N,N-Dimethylformamide	~ 1	~ 9
	Urea	1.5	4.3
	N-Methylpyrrolidin-2-one	4.4	10.1
Methanol (A) +	Formamide	2.3	0.5
DMSO (B)[h]	N-Methylformamide	1.9	0.5
	N,N-Dimethylformamide	0.8	0.5

The preferential solvation parameter $p < 1$ when solvent B is preferred. Where there are two entries in a column, the first pertains to water-rich, the second to organic-rich mixtures.
[a]GR Bebahani, D Dunnion, P Falvey, K Hickey, M Meade, Y McCarthy, MCR Symons, WE Waghorne. J Solution Chem 29:521, 2000.
[b]From infrared spectroscopy rather than calorimetry.
[c]M Meade, K Hickey, Y McCarthy, WE Waghorne, MCR Symons, PP Rastogi. J Chem Soc Faraday Trans 93:563, 1997.
[d]A Costigan, D Feakins, I McStravick, C O'Duinn, J Ryan, WE Waghorne. J Chem Soc Faraday Trans 87:2443, 1991.
[e]D Feakins, P Hogan, C O'Duinn, WE Waghorne. J Chem Soc Faraday Trans 88:423, 1992.
[f]G Carthy, D Feakins, C O'Duinn, WE Waghorne. J Chem Soc Faraday Trans 87:2447, 1991.
[g]Only in water-rich mixtures.
[h]D Feakins, CC O'Duinn, EW Waghorne. J Solution Chem 16:907, 1987.

standard molar enthalpy of solution of hydrophobic solutes (tetraalkylammonium halides) in such mixtures obeys the relationship:

$$\Delta_{sln}H^\circ = x_W \Delta_{sln}H^\circ_{(W)} + (1 - x_W)\Delta_{sln}H^\circ_{(DMF)} + (x_W^m - x_W)Hb \quad (5.54)$$

where Hb is the enthalpic effect of hydrophobic hydration and m is approximately the number of water molecules involved in this. Preferential solvation of solutes such as the polyols that have functional groups capable of accepting hydrogen bonds from the water and/or donating them to the water and DMF requires the addition of another term to Eq. (5.54), ΔH_{PS}. This can be written in terms of the average solvent replacement equilibrium constant, $K_N^{1/N}$, Eq. (5.7), as follows:

TABLE 5.3 Preferential Solvation and Enthalpies of Solution of Alkanols in Aqueous DMF Mixtures

Alkanol or Polyol	n_{OH}	m	$K_N^{1/N}$	$-Hb/\mathrm{kJ\,mol}^{-1}$
Methanol	1	8.02	1.88	9.12
Ethanol	1	7.7		13.4
1-Propanol	1	8.5		16.4
2-Propanol	1	7.2		18.8
1-Butanol	1	10.1		18.0
2-Methyl-1-propanol	1	9.3		18.6
2-Butanol	1	8.1		21.2
2-Methyl-2-propanol	1	7.0		24.1
1-Pentanol	1	11.2		19.3
2-Methyl-2-butanol	1	7.8		26.4
1,2-Ethanediol	2	6.83	1.99	11.5
Glycerol	3	8.08	1.78	11.3
meso-Erythritol	4	7.22	1.68	13.6
Xylitol	5	6.45	1.69	16.1
D-Arabinitol	5	6.55	1.73	16.2
L-Arabinitol	5	6.78	1.74	16.5
Ribitol	5	6.21	1.84	18.8
D-Sorbitol	6	5.99	1.73	20.3
D-Mannitol	6	5.41	1.77	21.2

Source: Ref. 54.

$$\Delta H_{PS} = NRT[(1 - x_W)/\{(1 - x_W) + x_W K_N^{1/N}\} - (1 - x_W) \ln K_N^{1/N}]$$
$$(5.55)$$

The results of the calculations with the combined Eqs. (5.54) and (5.55) for 10 alkanols and 9 polyols are shown in Table 5.3, assuming that $N = 3n_{OH}$, where n_{OH} is the number of hydroxyl groups in the polyol. It is seen that the values of $K_N^{1/N}$ vary little among the polyols, that m decreases moderately as n_{OH} increases, but the values of Hb are linear with this number: $Hb/\mathrm{kJ\,mol}^{-1} = -5.8 - 2.29\,n_{OH}$. However, Hb also increases linearly with the number of C–H bonds, which equals $n_{OH} + 2$, and this quantity appears to be responsible for the hydrophobic effect. Considerations similar to those involved in Eq. (5.54) have also been applied by Balk and Somsen [54] to alkanols, amines, and amides transferring into aqueous DMF.

5.3 ELECTROLYTE SOLUTES

5.3.1 Thermodynamic Studies

The preferential solvation of electrolytes that are practically completely dissociated into their constituent ions makes sense only for the preferential solvation of the individual ions because these species are those that are surrounded by the solvent molecules. Means must therefore be sought to define what is meant by the preferential solvation of ions and to relate this concept to measurable quantities. As for uncharged species, the preferential solvation of an ion by the components of a binary solvent mixture denotes the existence of a composition of the environment of the ion different from that of the bulk mixture. Electrolyte solutions differ from nonelectrolyte ones in that even at infinite dilution there are always particles of opposite and equivalent charges present in the solution. The use of spectroscopic information for the elucidation of the preferential solvation of ions does not suffer from this problem. The effects of the preferentially solvated cations and anions on measured spectroscopic quantities can generally be distinguished. Often, only the ions of one kind of sign have a measurable effect.

On the other hand, the use of thermodynamic data is handicapped by the inability of thermodynamics to deal with individual ions without the introduction of an extrathermodynamic assumption. Still, if a reasonable extrathermodynamic assumption is applied to measurable data, this permits the use of thermodynamically derived expressions for the investigation of the preferential solvation of ions. Such an assumption is the TATB one, stating that the cation and anion of tetraphenylarsonium tetraphenylborate are solvated equally in neat solvents [56]. The considerations leading to the TATB assumption should not lose their validity in solvent mixtures, as the large symmetrical ions with centrally hidden small charges of TATB should not be able to distinguish between the components of the mixtures. However, this is just an assumption, used in lieu of definite confirming data and in the absence of strong counter information.

A comprehensive critical review of the standard molar Gibbs energies of transfer of cations from water (A) to numerous aqueous cosolvent (B) mixtures, $\Delta_t G°$(cation, A $\rightarrow$ A + B) as a function of x_B was published by Kalidas et al. [57]. The data used were mainly from the solubilities of sparingly soluble salts and from the electromotive force measurements of suitable electrochemical cells. The splitting of the electrolyte data into the individual ionic contributions was by the TATB assumption, where made possible by the availability of the necessary data. Otherwise, other assumptions that were also deemed reliable had to be used. For many of the cation–cosolvent systems, values could be selected by the compilers from a combination of the reported data as best representing the "true" values of $\Delta_t G°(x_B)$. There were too few data for $\Delta_t G°$(anion, A → A + B) (with A =water) or $\Delta_t G°$(ion of any sign, A → A + B) (with A a nonaqueous solvent) reported by several independent sources to warrant a critical review. Subsequently, a comprehensive review of the standard molar enthalpies and entropies of transfer of salts and ions from water (A) to numerous aqueous cosolvent (B) mixtures, $\Delta_t H°$ or $\Delta_t S°$ (salt, A → A + B), was published by Hefter et al. [58]. The data were derived preferably from calorimetric measurements of the heats of solution, yielding $\Delta_t H°$, and from the temperature dependence of the electromotive forces of suitable cells, yielding $\Delta_t S°$. Because of the lack in many cases of data permitting the splitting into ionic contributions (by the TATB assumption) and of validating data from independent sources, this review dealt only fragmentarily with individual ions or suggested selected values on critical examination of most of the reported $\Delta_t H°$ data.

The dissociation equilibrium constants of weak acids in aqueous cosolvent mixtures were studied in a series of reports by Ohtaki and coworkers [59–61]. The proton transfer equilibrium between the hydronium ion H_3O^+ and the protonated cosolvent plays an important role in the variation of the pK values. Hence, the autoprotolysis constants of the mixtures had to be studied too. The results for aqueous alkanols are shown in Fig. 5.6 [62]. Barbosa and coworkers [63] used the QLQC method (see the following) to obtain the preferential solvation of the ions of carboxylic and other weak acids and explain the variation of their dissociation constants measured in mixed aqueous solvents. Such solvents are used as mobile phases in reversed-phase liquid chromatography and the acids are used for the adjustment of the pH. Aqueous acetonitrile and tetrahydrofuran have been subjected to such studies of the dissociation constants of acetic, tartaric, phthalic, citric, boric, and phosphoric acid. The preferential solvation parameters $\delta x_{MeCN(H^+)}$ are shown in Fig. 5.7, as are the pH values measured for the standard acetate buffer in the mixtures relative to the value in water.

The problematic use of thermodynamic data for the study of the preferential solvation of ions in binary solvent mixtures has been

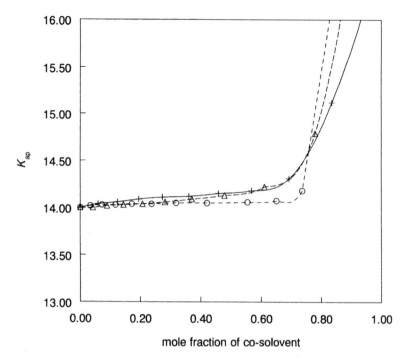

FIGURE 5.6 The autoprotolysis constants, $K_{ap}(x_B)$, of aqueous B: methanol
(-Δ- -Δ-), ethanol (+ + +), and 1-propanol (-o- - o-). (From Ref. 62.)

pointed out with reference to the IKBI method. Consider, for simplicity,
a symmetrical electrolyte, where the charges of the cation (subscript +)
and anion (subscript −) are equal but of opposite sign, so that
$z_+ + z_- = 0$, and their number concentrations are equal, $\rho_+ = \rho_-$.
The condition of electroneutrality of the solution as a whole requires
that on integration to infinity, the Kirkwood–Buff integrals defined by Eq.
(5.24) obey

$$z_+\rho_+ \int_0^\infty [g_{A+}(r) - 1]4\pi r^2 \, dr + z_-\rho_- \int_0^\infty [g_{A-}(r) - 1]4\pi r^2 \, dr = 0$$

$$(5.56)$$

since the counterion of an ion of either sign must be located somewhere in
the solution, be it dilute to infinity. From Eq. (5.56) follows that $G_{A+} =
G_{A-}$ and similarly $G_{B+} = G_{B-}$, or that:

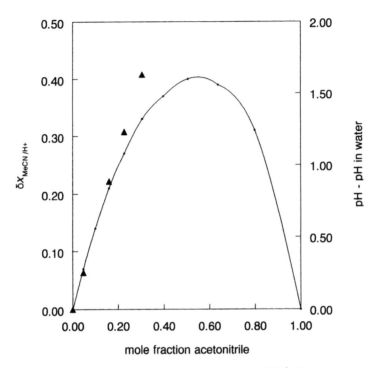

FIGURE 5.7 The preferential solvation parameter of H^+ in aqueous acetonitrile according to the QLQC method (left-hand ordinate) and the difference between the pH of the standard acetate buffer in these mixtures and that in water (right-hand ordinate, ▲) at 298.15 K. (From Ref. 63.)

$$G_{B+} - G_{A+} = G_{B-} - G_{A-} \tag{5.57}$$

or that the preferential solvation of the cations equals that of the anions, according to Eq. (5.28). This situation is certainly not the case in general. The problem rests with Eq. (5.25), in which the approximation of taking the integral to infinity instead of to the correlation radius, R_c, which is readily applicable to uncharged species, cannot be made for charged species. The correlation radius can be chosen large enough to include the entire region of space in which the ion has an effect on its surroundings but sufficiently small to exclude the counterion, if the dilution of the electrolyte is very high. Hence, local affinities of the cation and anion to the two solvents do have different magnitudes. These quantities, for instance,

$$G_{A+}^{L} = \int_{0}^{Rc} [g_{A+}(r) - 1]4\pi r^2 \, dr \qquad (5.58)$$

for solvent A in the surroundings of the cations, and similarly for B and for the surroundings of the anions, are not the same for ions of different kinds of charge. The definition in Eq. (5.57) is consistent with the condition that

$$G_{A+}^{L} + G_{A-}^{L} = 2G_{A+} = 2G_{A-} \qquad (5.59)$$

shown by Hall [64] to be valid.

Consider now a reference electrolyte with large cations and anions, similar in all respects except for the sign of the charge. The solvation of such ions will be rather weak in the first place and should not exhibit different preferences for the two solvents. Such a reference electrolyte may be the tetraphenylarsonium tetraphenylborate (TATB) mentioned earlier (or the corresponding phosphonium salt). Still, the ions of TATB may be preferentially solvated, although weakly, by one of the components of the solvent mixture. The extrathermodynamic assumption consists, therefore, in taking for the reference electrolyte that $G_{A+}^{L} = G_{A-}^{L}$ and necessarily also $G_{B+}^{L} = G_{B-}^{L}$ (but $G_{A+}^{L} \neq G_{B+}^{L}$). For this electrolyte $G_{A+}^{L} = G_{A+}$, and the latter, having infinity as the upper limit of integration, can be obtained from thermodynamic data; see Eq. (5.33). The derivative of the standard molar Gibbs energy of transfer then pertains to either of the reference ions, and in view of the above, $\partial \Delta_{tr} G_{+}^{\circ} / \partial x_{B} = \partial \Delta_{tr} G_{-}^{\circ} / \partial x_{B}$ for the reference electrolyte [56]. We can, therefore, employ the experimental data for the transfer of the reference electrolyte toward the calculation of the affinity of the counterion (say, the anion) in an electrolyte involving only one of the ions of the reference electrolyte (say, the cation). This, in turn, can be employed to establish the affinities of other ions and provide a complete scale of preferential solvation of ions in the mixtures of the solvents A and B.

The QLQC method considers only the first solvation shell, hence can inherently deal with individual ions, because no integration to infinite distance is required. Therefore, the presence of the counterion somewhere in the solution at infinite dilution makes no difference. Still, for dealing with the preferential solvation of an individual ion I, its standard molar Gibbs energy of transfer, $\Delta_{t} G^{\circ}(I, R \rightarrow A + B)$ from a reference solvent R (which may also be either A or B) is required, and this, again, involves an extrathermodynamic assumption, for which the TATB assumption can serve well enough. However, if the $\Delta_{t} G^{\circ}$ for the individual ion are not available, an average value for cation and anion is obtained from the $\Delta_t G°$ of the entire electrolyte. There may be cases where this still conveys information of interest, if there are reasons to ignore the preferential solvation of either the cation or the anion and concentrate attention on the ion of the other kind of charge. References for the applications so far of these two methods, IKBI and QLQC, are summarized in Table 5.4. If both the cation and the anion prefer the same solvent in the binary solvent mixture, *homosolvation* is said to take place, but this is the

TABLE 5.4 Applications of the IKBI and QLQC Methods to the Elucidation of the Preferential Solvation of Ions in Binary Solvent Mixtures

			Reference	
Solvent A	Solvent B	Ions	IKBI	QLQC
Water	MeOH	Na^+, Cl^-	a, b	c, a
Water	EtOH	K^+, Cl^-	a	c, a
Water	THF	H^+, $MeCO_2^-$		m
Water	DMSO	Li^+, Na^+, Cs^+, Ag^+, Cl^-, Br^-, I^-	a, d	e, a, k
Water	MeCN	H^+, Na^+, Ag^+, $Cu(I)$, Cu^{2+}, Cl^-, $MeCO_2^-$	a, e	c, e, f, g, l
Water	Pyridine	Zn^{2+}, F^-, IO_3^-, $PhCO_2^-$		j
MeCN	DMSO	Na^+, K^+, F^-, Cl^-, Br^-, I^-		c, h, k
MeCN	DMF	K^+, F^-, Cl^-, Br^-		h
DMF	DMSO	K^+, F^-, Cl^-, Br^-		h
EtOH	FA	K^+, F^-, Cl^-, Br^-		i
EtOH	NMF	K^+, F^-, Cl^-, Br^-		i
FA	NMF	K^+, F^-, Cl^-, Br^-		i

[a] Y Marcus. J Chem Soc Faraday Trans 1 85:3019, 1989.
[b] K E Newman. J Chem Soc Faraday Trans 1 84:1387, 1988.
[c] Y Marcus. Aust J Chem 36:1719, 1983.
[d] Y Marcus. Pure Appl Chem 62:2069, 1990.
[e] Y Marcus. J Chem Soc Faraday Trans 1 84:1465, 1988.
[f] Y Marcus. J Chem Soc Faraday Trans 87:1843, 1991.
[g] Y Marcus. J Chem Soc Dalton Trans 1991:2265.
[h] A-KS Labban, Y Marcus. J Solution Chem 20:221, 1991.
[i] A-KS Labban, Y Marcus. J Solution Chem 26:1, 1997.
[j] AV Varghese, C Kalidas. Z Naturforsch 46A:703, 1991.
[k] R Jellema, J Bulthuis, G van Zwan, G Somsen. J Chem Soc Faraday Trans 1 92:2569, 1996.
[l] J Barbosa, V SanzNebot. J Chem Soc Faraday Trans 90:3287, 1994; J Barbosa, D Barron, R Berges, V SanzNebot, I Toro. J Chem Soc Faraday Trans 93:1915, 1997.
[m] D Barron, S Buti, M Ruiz, J Barbosa. Phys Chem Chem Phys 1:295, 1999; Polyhedron 18:3281, 1999.

less common situation. The more common one, *heterosolvation*, takes place when the cation prefers one solvent and the anion prefers the other. This situation is exemplified by silver sulfate in aqueous acetonitrile, where the silver cation strongly prefers the acetonitrile and the sulfate anion strongly prefers the water [65].

The quantitative application of the QLQC method for the study of the preferential solvation of ions in binary solvent mixtures is illustrated by the work of Labban and Marcus [66,67], involving nonaqueous solvent mixtures. The solubility of potassium fluoride, chloride, and bromide in the solvents and solvent mixtures studied is sufficiently low and the relative permittivities are sufficiently high that corrections for activity coefficients can be applied and ion pairing is inconsequential. The standard molar Gibbs energy of solution can, therefore, be calculated from the solubilities at 298.15 K, as can the $\Delta_t G^\circ$ values for the transfer of the respective electrolyte from one neat solvent to another or the binary mixture. In the case of mixtures of the dipolar aprotic solvents DMF and DMSO, linear dependence of $\Delta_t G^\circ$ on the mole fraction composition of the mixtures is observed for all three salts. No preferential solvation of the ions of these salts takes place in such mixtures because the exact compensation of the preference of, say, the cation for DMSO and of the anion for DMF (or vice versa) for all three salts is unlikely to occur. However, for binary mixtures containing either of these two solvents (A) and MeCN (B), nonlinear curves of $\Delta_t G^\circ(x_B)$ are found. The agreement of the results calculated by means of the QLQC theory using $Z = 4$, Eq. (5.50), with the experimental values of $\Delta_t G^\circ(x_B)$ is shown in Fig. 5.8. In all these systems it is the potassium ion that is primarily solvated by the solvents, as the halide anions have little affinity for them [66]. An extension of these studies to mixtures of three protic nonaqueous solvents shows negligible preferential solvation of the ions of the salts in mixtures of formamide and NMF but does show preferences for the amides (A) over ethanol (B) in mixtures of either amide or ethanol. Again, the data are well represented by the QLQC method, Eq. (5.50), using $Z = 4$, Fig. 5.8 [67].

The QLQC method can explicitly consider the replacement of one solvent in the surroundings of the ion by the other as an equilibrium process [44]. In A-rich mixtures, the addition of component B causes the partial replacement according to Eq. (5.6a) (with the ion I being the solute S and the coordination number N being replaced by the lattice parameter Z). The equilibrium constant is

$$K_{A/B1} = x_{Z-1}x_A/x_Zx_B \qquad (5.60a)$$

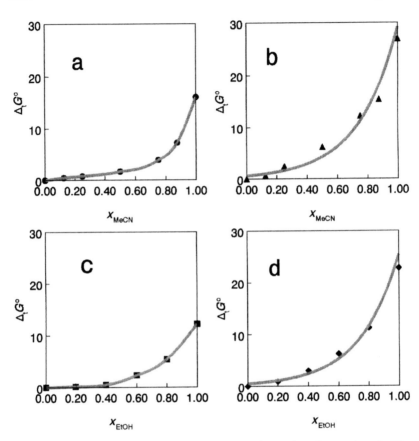

FIGURE 5.8 The standard molar Gibbs energies of transfer of potassium halides into binary solvent mixtures. Experimental data: (a) transfer of KCl from DMF into DMF + MeCN; (b) transfer of KBr from DMSO into DMSO + MeCN; (c) transfer of KF from NMF into NMF + EtOH; (d) transfer of KBr from formamide into formamide + EtOH. Curves are calculated by means of the QLQC theory, Eq. (5.50), using $Z = 4$. (From Refs. 66 and 67.)

where the solvated species of the ion are denoted by their subscripts (x_Z for IA_Z, etc.). In mixtures rich in component B, the corresponding replacement equilibrium is (5.6c) with an equilibrium constant:

$$K_{A/BZ} = x_Z x_A / x_{Z-1} x_B \qquad (5.60b)$$

completing the replacement of solvent A with solvent B. The reverse of Eq. (5.6a) can be written for the first replacement of solvent B with A, the corresponding equilibrium constant being $K_{B/A1} = 1/K_{A/BZ}$. The equilibrium constants can be calculated from the limiting values of the preferential solvation parameters:

$$K_{A/B1} = \lim(x_B \rightarrow 0)[x_B + \delta x_{1B}]/x_B[1 - x_B - \delta x_{1B}] \tag{5.61a}$$

$$K_{A/BZ} = \lim(x_B \rightarrow 1)[(1 - x_B)(x_B + \delta x_{1B})]/[1 - x_B + \delta x_{1B}] \tag{5.61b}$$

Some values for the solvent replacement equilibrium constants and the composition of the solvation shell near the ion in aqueous cosolvent mixtures are shown in Table 5.5.

A more general treatment of the replacement of one solvent by the other in the solvation shell of an ion was presented by Cox, Parker, and Waghorne [68] and by Covington and Newman [69]. These authors considered the stepwise replacement of solvent A by solvent B, as in the coordination model for metal ion complexes with any ligand B, and up to the coordination number N (corresponding to Z in the QLQC method); see Eqs. (5.6a) to (5.6c). For the overall replacement of i A solvent molecules with i B molecules, the equilibrium constants are β_i. In the treatment of Cox et al. [68], volume fractions of the solvents were used instead of molar concentrations (or mole fractions):

$$\beta_i = (c_i/c_0)(\phi_B/\phi_A)^i \tag{5.62}$$

where the solvated ion species are denoted by the subscripts corresponding to the number of B molecules coordinated. Because the ions are at

TABLE 5.5 Replacement Equilibrium Constants and Compositions of the Solvation Shell at $x_B = 0.5$ (with the Lattice Parameter $Z = 6$) from the QLQC Method for Mixtures of Cosolvents (A) with water (B)[a]

Cosolvent (A)	Ion	K_{B1}	K_{BZ}	x_A^{Lz}	x_B^{Lz}
Dimethylsulfoxide	Na^+	0.47	0.87	3.7	2.3
	Ag^+	0.23	0.42	4.6	1.4
	Cl^-	2.85	5.3	1.2	4.8
Acetonitrile	Na^+	2.46	1.12	2.3	3.7
	Ag^+	0.68	0.31	4.1	1.9
	Cl^-	6.1	2.8	1.2	4.8

[a]Note that the designations A and B are opposite those generally used in this chapter.
Source: Ref. 44.

negligible concentration in the solvent mixture, the *ratios* of the species concentrations become independent of the concentration scale. Hence, mole or volume fractions can be used instead of the molar concentrations, c, of the solvated ion species in Eq. (5.62). The standard molar Gibbs energy of transfer of the ion from neat solvent A into the mixture is given by [68]

$$\Delta_t G^\circ (I, A \rightarrow A + B) = -RT \ln \phi_A - RT \ln \left[1 + \sum \beta_i (\phi_B / \phi_A)^i \right]$$

$$(5.63)$$

where the sum on the right-hand side extends from $i = 1$ to $i = N$ (in the original paper the sign before the $\sum$ in Eq. (5.63) was misprinted as $-$ instead of $+$). The number of B molecules in the coordination sphere is calculated in the same manner as is the average ligand number in solution coordination chemistry to give the local mole fraction of solvent B, $x_B^L = N_B / N$, as

$$N_B / N = \sum i \beta_i (\phi_B / \phi_A)^i / N \left[1 + \sum \beta_i (\phi_B / \phi_A)^i \right] \qquad (5.64)$$

The standard molar entropy of transfer of the ion is related to the change in the configurational entropy accompanying the replacement of one solvent with the other [51]:

$$S_{conf} = -n_A R \ln[(n_A / n) / x_A] - n_B R \ln[(n_B / n) / x_B] \qquad (5.65)$$

The standard molar enthalpy of transfer is then obtained from $\Delta_t H^\circ (I, A \rightarrow A + B) = \Delta_t G^\circ (I, A \rightarrow A + B) + T \Delta_t S^\circ (I, A \rightarrow A + B)$. These expected relationships should be valid for systems that adhere to this model (in terms of volume fractions of the component solvents rather than their activities), and such systems are the transfer of Na^+ and Ag^+ from propylene carbonate to its mixtures with dimethylsulfoxide [68]; see Fig. 5.9. In other systems deviations are found, but preferential solvation according to the model still plays the major role in determining the thermodynamic functions of transfer of electrolytes.

    This model was later modified to the *extended coordination model* (ECM), applied to the enthalpies of transfer of both nonelectrolytes (see earlier) and electrolytes. The deconvolution of Eq. (5.52) in terms of the two parameters p and $(\alpha n + \beta N)$ is difficult and therefore was applied [52] only for two extreme cases: where the preferential solvation parameter $p \approx 1$ and where $p \gg 1$. For lack of transfer enthalpy data for individual ions, it was applied to entire electrolytes (alkali metal halides). For nearly

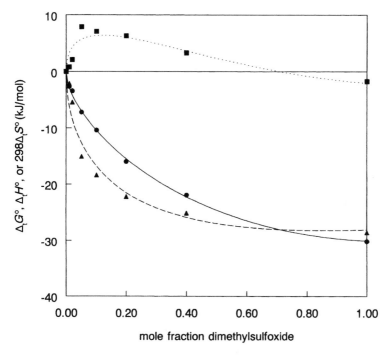

FIGURE 5.9 Transfer thermodynamics of NaCl from propylene carbonate to its mixture with dimethylsulfoxide at 298 K. Symbols: $(\bullet)\Delta_t G^\circ$, $(\blacktriangle)\Delta_t H^\circ$, and $(\blacksquare) - 298\Delta_t S^\circ$. (From Ref. 68.)

random solvation ($p \approx 1$), as for aqueous methanol, $\Delta_t H^\circ/x_B$ should vary linearly with $\Delta_t H^E/x_B \approx (x_A l_A + p x_B l_B)/x_B$. For strong preference for solvent B, as for aqueous acetonitrile, $\Delta_t H^\circ$ should vary linearly with l_B as long as $p x_B \gg x_A$; see Table 5.6. In general, p can be construed as the average equilibrium constant, $\beta_i^{I/i}$, and for the stepwise replacement of solvent A by solvent B around the ion according to Eqs. (5.6b) and (5.62).

The stepwise coordination solvation model was applied by Covington and Newman [69] in a somewhat different manner because these authors also considered electrostatic contributions to the Gibbs energy of transfer besides that arising from the preferential solvation. However, if the transfer is carried out under isodielectric conditions, these electrostatic contributions cancel out, and essentially Eq. (5.63) is recovered, although these authors preferred the use of the mole fractions

TABLE 5.6 Parameters for the Enthalpies of Transfer of Electrolytes, Eq. (5.52)

Electrolyte	$(\alpha n + \beta N)$ for MeOH + water $(p \approx 1)$	$(\alpha n + \beta N)$ for MeCN + water $(p \gg 1)$
LiCl	5.61	9.00
NaCl	6.10	9.71
KCl	5.66	12.90
RbCl	5.56	10.97
CsCl	5.88	12.98
NaBr	7.99	10.62
NaI	10.09	12.99

Source: Ref. 52.

of the solvents to their volume fractions. A statistical distribution of the equilibrium constants for the stepwise solvent replacement was assumed, yielding an average equilibrium constant:

$$K = [i\beta_i/(n + 1 - i)\beta_{i-1}]^n \tag{5.66}$$

for all steps i in Eq. (5.6b). The method was applied to NMR measurements, and it was assumed that the chemical shifts of the individual solvated species are proportional to the number of B molecules in the solvation shell, $\delta_i = (i/n)\delta_B$, where δ_B is the chemical shift in neat solvent B relative to solvent A. Therefore, the observed shift is

$$\delta = \delta_B \sum (ic_i)/n \sum c_i = \delta_B K^{1/n}(x_B/x_A)/[1 + K^{1/n}(x_B/x_A)] \tag{5.67}$$

The restriction to statistical distribution of the stepwise equilibrium constants can, however, be dispensed with [70,71]. The constancy of the coordination number N is also not an absolute requirement for this approach [71], as situations can be encountered where one A molecule in the solvation sphere is replaced by, say, two B molecules (so that $N_{neat\ B} = 2N_{neat\ A}$). However, the experimental data are seldom sufficiently accurate to require these complications. On the other hand, in the general case, the electrostatic effects on the Gibbs energy of transfer, Eq. (5.63), and that of the nonideality of the solvent mixture cannot be neglected. Covington and Newman [72] used NMR chemical shift measurements of monoatomic ions in various aqueous (A) solvent (B) mixtures to establish the average solvent replacement equilibrium constants, shown in Table 5.7. This model (called SSE), of solvent replacement around the solute, has been compared with the QLQC model of preferential solvation by Jellema et al.

TABLE 5.7 Solvation Replacement Equilibrium Constants $K^{1/N}$ for the Replacement of Solvent A by the Cosolvent B Around Ions, Obtained by NMR Chemical Shift Measurements

Solvent A	Cosolvent B	Ion	$K^{1/N}$
Water	Hydrogen peroxide	Li^+	0.34
		Na^+	0.97
		Rb^+	2.44
		Cs^+	3.0
		F^-	0.26
Water	Methanol	Na^+	0.61
		Rb^+	0.70
		Cs^+	0.76
		F^-	0.97
		Cl^-	0.70
Water	Dimethylsulfoxide	Li^+	1.32
		Cs^+	1.15
Water	Tetramethylenesulfone	Cs^+	2.7
1,2-Ethanediol	Acetonitrile	Na^+	72
		I^-	9.5

Source: Refs. 70 and 72.

[73], who demonstrated that the SSE model overcorrects for the solvent nonideality and nonstatistical distribution of the solvated species.

Jellema et al. [74] extended the QLQC approach to deal with the enthalpy of transfer of salts from solvent A to the mixed solvent A + B, utilizing the excess Gibbs energy expression and surface fractions θ and interaction parameters τ from the UNIQUAC approach (see Section 2.3). Their resulting expression is

$$\Delta_t H^\circ(\text{salt}, A \to A + B) = (1 - x_B)\Delta_t H^\circ(\text{salt}, B \to A) - H^{E*} + h$$

$$(5.68)$$

Here H^{E*} is the excess enthalpy of mixing according to the UNIQUAC expression:

$$H^{E*} = (ZRT/2)[x_B\theta_B(\theta_A + \theta_B\tau_A)^{-1}(T\partial\tau_A\partial T)$$
$$+ x_B\theta_A(\theta_B + \theta_A\tau_B)^{-1}(T\partial\tau_B\partial T)]$$

$$(5.69)$$

$$h = (ZRT/2)[(x_A \theta_B y \exp(\Delta) - x_B \theta_A)(\theta_A + \theta_B y \exp(\Delta))^{-1}]$$

$$\times [\theta_B (\theta_A + \theta_B \tau_A)^{-1} (T \partial \tau_A / \partial T) - \theta_A (\theta_B + \theta_A \tau_B)^{-1} (T \partial \tau_B / \partial T)$$

$$- \Delta_t H^\circ (\text{salt}, B \rightarrow A)] \tag{5.70}$$

where y and Δ are the quantities defined in Eqs. (5.48) and (5.49). The quantity h is able to explain the difference between solvent mixtures having positive H^E and G^E on the one hand (where H^E is often quite asymmetric with regard to the solvent composition) and mixtures having negative H^E and G^E on the other. For the calculation of the enthalpies of transfer, $\Delta_t H^\circ$, according to Eq. (5.68), values of τ and $\partial \tau_B / \partial T$ and $\theta(x)$ for both solvents from the UNIQUAC approach (Section 2.3) are required, as well as the transfer Gibbs energy and enthalpy of both ions of the salt (or the salt itself) from the one neat solvent to the other. The values of $\Delta_t H^\circ$ for sodium bromide and potassium iodide from water (A) to aqueous dimethylsulfoxide (B), where both H^E G^E are negative, calculated from Eq. (5.68) with $Z = 6$, agree well with the experimental values, being monotonous as expected. The calculated values of $\Delta_t H^\circ$ of these salts for transfer into aqueous acetonitrile (B), where both H^E and G^E are positive, have on the other hand a minimum (in the B-rich region), again in agreement with the experimental data [74].

Much earlier studies of the preferential solvation of ions in binary solvent mixtures neglected the direct preferential coordination of the ions by the components of the mixture and focused on the electrostatic effect. Scatchard [75] studied the sorting of mixed solvents by ions on this basis but concluded that the agreement of the derived expressions (for alkali halides in aqueous ethanol and dioxane mixtures) with experimental data is unsatisfactory. A later approach by Padova [76] arrived at the expression for the preferential solvation equilibrium quotient p:

$$RT \ln p = RT \ln[(x_B^L / x_A^L)/(x_B / x_A)]$$

$$= -\frac{1}{2} \int_{r_e}^{\infty} r^2 dr \int_0^{E_r} (\partial \mu_A / \partial \ln x_A)^{-1}$$

$$\times [(\partial \varepsilon_d / \partial n_A)_{n_B} - (\partial \varepsilon_d / \partial n_B)_{n_A}] d(E^2) \tag{5.71}$$

where the partial derivatives are at constant temperature, pressure, and electric field; r_e is the effective radius of the ion; E_r is the field at the distance r from the center of the ion; ε_d is the differential permittivity of the solvent mixture; and n_A and n_B are the numbers of moles of the components in the solvent mixture. The right-hand side of Eq. (5.71) could not be evaluated directly but could be shown to be given by (the electrostatic contribution to) the difference in the solvation Gibbs energies of the ion in the two neat solvents, hence by the standard molar Gibbs energy of transfer of the ion between them.

5.3.2 Spectroscopic and Structural Studies

Information concerning the preferential solvation of ions (or electrolytes) can, of course, also be obtained from nonthermodynamic information [77], such as spectroscopy, X-ray diffraction, computer simulations, and kinetic or transport properties. Following are illustrative and far from exhaustive examples that show the scope of the use of spectroscopic measurements for the study of preferential solvation of ions.

Earlier studies of, say, the light absorption in the UV region provided qualitative information concerning the preferential solvation of anions, e.g., iodide, by the protic constituents of the mixed solvent [77]. More quantitative results were subsequently obtained from the light absorption measurements. For instance, near-infrared measurements of the $v_2 + v_3$ combination band of water at $1.954 \, \mu m$ for low concentrations ($< 0.6 \, mol \, L^{-1}$) of water in DMSO containing the strong trifluorosulfonic acid [78] showed the preferential protonation of the water: $K(DMSOH^+ + H_2O \rightleftharpoons DMSO + H_3O^+) = 6.6$. Tanaka et al. [79] used the luminescence lifetimes τ of europium(III) ions or rather the changes Δk (relative to the corresponding deuterated solvent) in their reciprocals, $k = 1/\tau$. Thus, $\Delta k = x_{AS}^L \Delta k_A + x_{BS}^L \Delta k_B$ in the mixture and their preferential solvation parameter is $P_S = K_{AB/S}$. In aqueous acetone, acetonitrile, and dioxane, water was absolutely preferred even at as low water contents as $x_{water} = 0.1$. In aqueous methanol and dimethylsulfoxide, water was still preferred near the europium ions but to a limited extent. In aqueous N, N-dimethylformamide it is the amide that is preferred near the solute when its mole fraction is below 0.7, whereas water is preferred in the more amide-rich mixtures.

Kleeberg [80] studied the preferential solvation of copper(II) perchlorate in aqueous acetonitrile mixtures. He used the IR wavenumbers of the O–H versus O–D stretching frequencies in the water, the C–N stretching frequency in acetonitrile, and the Cl–O stretching frequency in the perchlorate anion, as well as the intensity of light absorption by the copper(II) ion in the visible spectrum. A new subband at $2328 \, cm^{-1}$ of the acetonitrile, not seen in the absence of the copper(II), characterizes the molecules of this solvent polarized by the ions and was used to determine the preferential solvation, that is, the number of acetonitrile molecules in the first solvation shell of the copper ion, $Nx^L_{MeCN(Cu)}$. The ions preferred water over the acetonitrile in their vicinities in dilute acetonitrile mixtures, but $Nx^L_{MeCN/Cu}$ rose steeply at bulk $x_{MeCN} > 0.6$ and a total coordination number of $N = 6$ around the copper ions was established [80]. The C – N stretching frequency in acetonitrile (as well as the N–C–O deformation frequency in DMF) was used to study the preferential solvation of Li^+ ions in mixtures of acetonitrile with DMF. The intensities of the deconvoluted free and solvating solvent species of both components were derived, and preferential solvation by the DMF was established [81].

The NMR signals from ^{63}Cu and ^{65}Cu were measured in copper(I) perchlorate solutions in aqueous acetonitrile by Gill et al. [82]. Quadrupolar line broadening permitted the detection of the signals only from the symmetrical $Cu(MeCN)_4^+$ ions, but not when water replaced some of the acetonitrile, and no quantitative results could be obtained from this technique. Still, preferential solvation of Cu^+ by the acetonitrile was evident. In mixtures of acetonitrile with methanol, DMF, and trimethylphosphate the Cu^+ was again preferentially solvated by the acetonitrile, but when the cosolvent was pyridine or trimethylphosphite the latter solvents were preferred [82]. Other early studies of preferential solvation of ions by NMR were the chemical shift measurements by Fratiello et al. [83–86], which also yielded mainly qualitative results. Preferential hydration of the cations was indicated in aqueous dioxane, tetrahydrofuran, or pyridine mixtures. Beryllium and nickel ions, however, showed competition of the water and pyridine for sites in the solvation shell. However, NMR measurements have since been used by numerous authors in other systems to study preferential solvation quantitatively. Thus, cooling the solution down can slow the solvent exchange sufficiently to obtain separate signals from the two solventt components in the solvation shell [77,87].

Finter and Hertz [88], Holz [89], and Holz et al. [90] measured the NMR spin-lattice relaxation of ionic nuclei in protonic and deuterated solvent mixtures. In aqueous N-methylformamide (NMF) mixtures $^{127}I^-$ and $^{23}Na^+$ ions were used. The NMF was found to be preferred by the iodide ion but water by the sodium ion, resulting in *heterosolvation*. When aqueous formamide (FA) was studied [88], the sodium ions preferred the FA, as did the iodide ion, resulting in *homosolvation*. More complicated behavior was found when sodium bromide, rather than the iodide, was used in aqueous NMF: the sodium ions preferred the water only when its mole fraction did not exceed 0.6, otherwise they joined the bromide ion and preferred the NMF. Sodium ions preferred dimethylsulfoxide over water [89], but water was preferred over acetonitrile by this ion over the entire concentration range. However, for hexamethyl phosphoric tri-amide (HMPT) it showed little preference when its concentration in the aqueous mixture was rather low or rather high, but water was preferred at intermediate concentrations [90].

Covington and Dunn [70] measured the ^{23}Na NMR chemical shifts and the I^- charge transfer band of sodium and iodide in mixtures of acetonitrile with 1,2-ethanediol and 1,2-propanediol. Both ions preferred the glycols over the nitrile. Proton NMR chemical shifts of acetonitrile were used by Schneider [65] in aqueous acetonitrile solutions of silver nitrate to obtain the number of acetonitrile molecules solvating the silver ions, $Nx^L_{MeCN/Ag}$. The equilibrium constants for the solvent exchange between water and DMSO, MeCN, and propylene carbonate (PC) in the solvation shells of ions were determined from the proton NMR chemical shifts of low concentrations of water [78] at 303.15 K. The resulting $K =$ [hydrated ion]/[organic-solvated ion]a_{water} (the activity of the organic solvent being taken as unity) are shown in Table 5.8, demonstrating the preferential hydration of the ions in all three solvents studied (except for K^+ in PC). In the case of silver ions, however, Ag^+ is preferentially hydrated at high water contents, but it is solvated by DMSO when x_{water} < 0.33 [91]. Because of the slow exchange of the solvents in the sol-vation shell of beryllium ions, four separate ^{31}P NMR signals were obtained [92,93] for beryllium nitrate solutions (>1 mol L^{-1}) in aqueous hexamethyl phosphoric triamide, HMPT (up to $x_{HMPT} = 0.493$). The individual solvent exchange constants, Eq. (5.6), for the species

TABLE 5.8 Equilibrium Constants for the Solvent Exchange Reaction in the Solvation Shells of Ions: $K =$[Hydrated Ion]/[Cosolvent-Solvated Ion]a_{water}, Obtained by Water-Proton NMR Chemical Shift Measurements

Ion	K in DMSO	K in MeCN	K in PC
H$^+$	16	≫100	≫100
Li$^+$	85	13	7.8
Na$^+$	45	8.4	1.7
K$^+$		4.2	0.5
Cl$^-$	98	35	7.5

Source: Ref. 78.

$Be(H_2O)_{4-i}$ $(HMPT)_i$ could then be calculated. These constants are considerably higher than in the corresponding system with DMSO. For magnesium perchlorate solutions in aqueous acetone or acetonitrile the solvent exchange is fast, hence the solvent replacement equilibrium constants, Eq. (5.6), had to be evaluated from the line widths of the ^{25}Mg NMR signals, corrected for the solution viscosity. Because water is strongly preferred over the cosolvent, the notation water $= B$ and cosolvent $= A$ is used here, contrary to the usage elsewhere in this chapter. The values of $K_{A/Bi}$ (with $N = 6$) vary from $K_{A/B1} > 10^5$ to $K_{A/B6} = 2.5$ for both cosolvents, being slightly larger in the acetonitrile than in the acetone system for the intermediate constants [94].

The information obtained on preferential solvation of ions from kinetic and transport measurements is essentially only qualitative. Kinetic studies of S_N2 reactions, such as replacement of a halogen atom on an alkyl halide by another, showed that protic cosolvents affect the reaction rate by the ability to solvate the incoming anion. The rate-retarding sequence for the reaction $CH_3I + Cl^- \rightarrow CH_3Cl + I^-$ in dimethylacetamide is $PhOH > PhSH > MeOH > PhNH_2 > H_2O$, where all these cosolvents are at the same molar concentration 1.0 mol L^{-1}[95]. When the concentration of the cosolvent, methanol, in the dimethylacetamide was increased, the rate was diminished continuously, without showing the formation of discrete methanol solvates of the incoming ion. However, the data clearly indicated preferential solvation of the halide ions by the protic cosolvents. In S_N1 solvolysis reactions, e.g., of t-butyl chloride, the Gibbs energy of activation became less negative in aqueous methanol, as the methanol content increased; i.e., water preferentially solvated the outgoing chloride anion [96]. However, these and similar pieces of information are all qualitative, confirming the preferential solvation by the binary solvent component that is better able to donate hydrogen bonds to the anions, but do not provide quantitative measures of the extent of this preferential solvation. A quantitative treatment based on the Kirkwood–Buff approach to preferential solvation and on transition state theory was proposed by Blandamer and coworkers [97]. They defined the *affinity change*, A_{kA}, of the reactant for solvent A on going from the initial state to the transition state at a mole fraction x_B of the cosolvent:

$$A_{kA}(x_B) = (V_B x_B / RTQ)[\partial \Delta G^{\neq}(x_B)/\partial x_B] + \Delta V^{\neq}(x_B) \qquad (5.72)$$

where Q is the quantity defined in Eq. (5.31), $\Delta G^{\neq}$ is the Gibbs energy of activation, and $\Delta V^{\neq}$ is the volume of activation for the reaction. Then the derivative of the logarithm of the ratio of rate constants in the presence and absence of the cosolvent B becomes

$$\partial \ln[k(x_B)/k(0)] = -(Q/V_B x_B)[A_{kA}(x_B) - \Delta V^{\neq}(x_B)]\partial x_B \qquad (5.73)$$

This expression describes the dependence of the kinetics of the reaction on the composition of the mixed solvent.

Transport properties of electrolytes in binary solvent mixtures also provide mainly qualitative information concerning preferential solvation of the ions. The classical Hittorff method for the measurement of the transference numbers of the ions also produced information concerning the amounts of the solvent components A and B carried along with the ions, as shown by Strelow and Koepp [98]. The relevant quantity is Δ, when a current i is passed for a time t, transporting a mass m of solvent with the transferred ions:

$$\Delta_B = Fm(itM)^{-1}[x_B(t) - x_B(0)] \qquad (5.74)$$

where F is the Faraday constant and M is the mean molar mass of the solvent mixture transferred [98]. Otherwise, the quantity Δ_B can be obtained from the electromotive force (EMF) of a cell with transference, but this requires the derivative of the activity coefficient of B with respect to its mole fraction in the bulk solvent [99]. The difference of mole fractions of component B per mole of ions transferred (F/it) is a function of four solvation numbers n:

$$\Delta_B = (x_A n_{B+} - x_B n_{A+})t_+ - (x_A n_{B-} - x_B n_{A-})t_- \qquad (5.75)$$

where subscripts $+$ and $-$ denote the cation and anion and the $t_\pm$ are the transference numbers of the ions (for symmetrical electrolytes, otherwise the $t_\pm$ are divided by the charges of the ions). For heterosolvated electrolytes n_{B+} and n_{A-} may be large whereas n_{A+} and n_{B-} may be small, so that large positive Δ_B results. Otherwise, only small Δ_B values are obtained, from which little information concerning the preferential solvation of the ions can be deduced. Some examples of the variation of Δ_B with the solvent composition for heterosolvated electrolytes are shown in Table 5.9. Due to the dependence of $\Delta_B(x_B)$ in general on four independent solvation numbers n as in Eq. (5.75), no quantitative information can be derived concerning the preferential solvation of the individual ions of the electrolyte, except for the qualitative confirmation of the heterosolvation in the system studied.

Little direct information from X-ray diffraction on the preferential solvation of ions in binary electrolyte mixtures can be obtained. This is

TABLE 5.9 The Difference Δ_B of Solvent Transport with Electrolytes in Binary Solvent Mixtures A + B at 30°C as a Function of the Mole Fraction of B, x_B

AgNO₃ A=water, B=MeCN		AgIO₃ A=EtOH, B=MeCN		ZnF₂ A=water, B=pyridine		Cu(IO₃)₂ A=EtOH, B=DMSO	
$100x_B$	Δ_B	$100x_B$	Δ_B	$100x_B$	Δ_B	$100x_B$	Δ_B
0.55	0.16	5	0.48	15	0.02	15	0.29
14.5	0.78	15	1.1	25	0.68	25	0.89
40.0	2.04	25	1.2	35	0.91	35	1.27
76.0	1.46	35	2.0	45	2.2	45	1.20
94.2	0.70	45	2.1	55	2.2	55	0.79
		55	2.4	65	0.98	65	0.53
		65	1.3	75	0.91	75	0.43
		75	0.8	85	0.23	85	0.12
		85	0.24			95	0.34

Source: Ref. 99 (except for AgNO₃, 25°C, from Ref. 98).

true in particular if the solvent components bind to the ions through the same kind of atom, e.g., the oxygen atom in mixtures of water and methanol. Electrolytes have been studied in these neat solvents quite extensively, but in a study of an electrolyte, namely magnesium chloride, in these mixtures no conclusions concerning preferential solvation could be drawn by Radnai et al. [100]. The preferential hydration of the ions of sodium chloride in aqueous acetonitrile mixtures, studied by Takamuku et al. [101] using large-angle X-ray and small-angle neutron diffraction, was shown to be the cause of the induced separation of the mixture into two liquid phases on addition of the salt.

On the contrary, computer simulations appear to be promising in this respect. With this method it is possible to focus on an individual ion, and its interaction potentials with the components of the mixed solvents can be obtained from computer simulations of the ion in the neat solvents. Thus, in a study of a solute ion I in the binary mixture A + B, there are five interaction potentials that need to be known: I-A, I-B, A-A, A-B, and B-B. A Monte Carlo computer simulation experiment by Galera et al. [102] regarding a chloride anion in an equimolar mixture of water and methanol clearly showed the preferential hydration of the chloride anion. There are on the average six water molecules and a single methanol molecule in the vicinity of this anion. In a molecular dynamics study of the solvation of a sodium or of a chloride ion in water (205–215 molecules) containing a single HMPT molecule Kessler et al. [103] showed preferential solvation by the amide of the cation but not of the anion. That is, the HMPT molecule was found, on the average, in the vicinity of the sodium ion but did not compete with the water around the chloride ion. More recently, molecular dynamics was applied to the study of sodium and chloride ions in aqueous methanol by Noskov et al. [104] and in aqueous formamide by Puhovski and Rode [105].

5.4 PREFERENTIAL SOLVATION OF LARGE MOLECULES

Large molecules, such as polymers or biochemical multifunctional molecules, are often dissolved in binary solvent mixtures. The purpose is, generally, to enhance the solubility, if the solute is insufficiently soluble in either of the constituents of the mixture. In such situations, the two kinds of solvent molecules can arrange themselves differently around different parts of these large molecules. In other cases, a so-called nonsolvent, in which the large solute molecules are practically insoluble, is added to a good solvent in order to precipitate it. It is therefore of interest to explore the preferential solvation of such solutes. The group contribution concept is of use in such cases, where each functional group of the large molecule (together, if necessary, with the attached part of the "skeleton" of the large molecule) is treated as if it were a "small" molecule solvated in the solvent mixture. Mutual interactions of such groups are either ignored or taken into account in more sophisticated investigations. The solubility of the solute, as also other partial molar thermodynamic functions, can be derived from the Gibbs energy of solvation of the solute, so the solvent composition dependence of this quantity is to be described in the following. It is tacitly assumed that internal degrees of freedom of the solute as well as its conformation are not affected by the gradual change of the solvent composition.

Ben-Naim [106] extended the inverse Kirkwood–Buff integral method for the elucidation of preferential solvation to cover also large solute molecules S in binary solvent mixtures, A + B. Equation (5.29), or equivalently Eq. (5.30), was modified for the present purpose as follows. The total solvation Gibbs energy of the solute is divided [107] to include contributions from its "hard" and "soft" parts (superscripts H, S) as well as from "dangling" functional groups or side chains i extending from the hard/soft skeleton of the solute, Fig. 5.10a. The hard part is due to the interaction potential at distances from the centers of the atoms constituting the backbone of the solute, below which the solute can be considered as interacting with a hard sphere type of potential. The soft part is then due to the residual interaction potential. Thus:

$$\Delta_{solv} G_S^o = \Delta_{solv} G_S^{oH,S} + \sum_i \Delta_{solv} G_{Si}^o \qquad (5.76)$$

where $\Delta_{solv} G_S^{oH,S}$ is the part of the total solvation Gibbs energy where the soft part is added, given the hard part, and $\sum_i \Delta_{solv} G_{Si}^o$ is the part due to the functional groups or side chains being added, given that $\Delta_{solv} G_S^{oH,S}$ has already been taken into account. These added parts can be termed *conditional contributions* to the solvation Gibbs energy [107].

Corresponding to this division, the affinity of the solute to the solvent A can be written as:

$$G_{AS} = G_{AS}^{H,S} + \sum_i G_{ASi} \qquad (5.77)$$

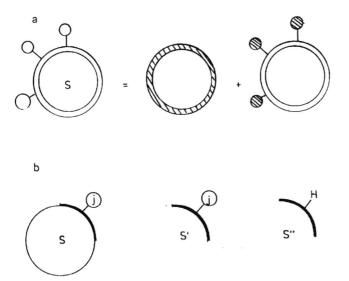

FIGURE 5.10 (a) A large molecule, split into the "hard" and "soft" (hatched) backbone parts and those from the functional groups (dangling circles). (b) A model, S', for solute S, having a functional group j with a part of the backbone attached to it, and its analog S'' with the functional group j replaced by a hydrogen atom (constituting only part of the backbone). (From Ref. 106.)

and similarly for G_{BS} (Note that here G is the affinity, obtained from the Kirkwood–Buff integral, not the Gibbs energy as previously. This confusing use of symbols is deplorable, but the symbols have been too deeply rooted in usage for them to be modified.) The quantity $\rho_A G_{AS}$ is the change in the number of A molecules in the correlation volume of S produced by placing it at some fixed position in the solvent mixture. Then $\rho_A G_{AS}^{H,S}$ is the corresponding change due to the placement of only the hard and soft parts of the solute, and $\rho_A \sum_i G_{ASi}$ is the *additional change* in the number of A molecules produced when the functional groups or side chains are added to the skeleton of the solute, already placed in the solvent mixture. At this stage, the effects of the various functional groups or side chains are thought of as affecting the solvent distribution independently. Then Eq. (5.30) can be written as

$$\lim(x_S \to 0)[\partial \Delta_{\text{solv}} G_S^O / \partial x_B]_{T,P} = C\Big[(G_{AS}^{H,S} - G_{BS}^{H,S})$$
$$+ \Big(\sum_i G_{ASi} - \sum_i G_{BSi}\Big)\Big] \quad (5.78)$$

where C is an appropriate constant. The main point in Eq. (5.78) is that the total preferential solvation effect has been separated into the several (conditional) contributions to it.

The way to proceed from here is to obtain results for model compounds S' that comprise only one functional group or side chain, j, and the part of the backbone attached to it, Fig. 5.10b. For such a model compound, Eq. (5.77) is rewritten as

$$\Delta_{solv} G^o_{S'} = \Delta_{solv} G^{oH,S}_{S'} + \Delta_{solv} G^o_{S'j} \tag{5.79}$$

Furthermore, a second model compound, S'', is used where the particular functional group or side chain j is replaced by a hydrogen atom, so that it comprises only the backbone part of S'. Hence the (conditional) solvation Gibbs energy of the functional group or side chain is obtained as $\Delta_{solv} G^o_{S'j} = \Delta_{solv} G^o_{S'} - \Delta_{solv} G^o_{S''}$, the two quantities on the right-hand side being measurable. Therefore, their measurement as a function of x_B in the solvent mixture provides the effect of the particular functional group or side chain j on the preferential solvation according to Eqs. (5.30) and (5.78). The hard and soft parts of the solvation Gibbs energy can be estimated from the solubilities of appropriate model compounds with structures similar to the backbone of the large molecule (e.g., alkanes for hydrocarbon polymers, oligopeptides for proteins). The relative volumes (surface areas, number of monomers) of the actual large molecule and the model compound must, of course, be taken into account. Otherwise, this part of the solvation Gibbs energy can be calculated from the work of cavity creation for the molecular skeleton, using the scaled particle theory [108]:

$$\Delta G_{cav} = RT\{-\ln(1-y) + 3y(1-y)^{-1}(\sigma_S/\sigma_{AB})$$
$$+ [3y(1-y)^{-1}(9/2)y^2(1-y)^{-2}](\sigma_S/\sigma_{AB})^2\}$$
$$+ PV_{AB}y(\sigma_S/\sigma_{AB})^3 \tag{5.80}$$

where $y = (\pi/6)N_{Av}\sigma^3_{AB}/V_{AB}$ is the solvent packing fraction, σ is the molecular diameter, and subscript AB denotes the average value for the binary solvent mixture.

In the general case, however, the effects of the functional groups or side chains are not independent of each other. Therefore, it is necessary to use data for several model compounds that have these molecular parts at different distances from each other to estimate their mutual effects. Another complication is that in actual cases the conformation of the large solute (e.g., polymer or protein) does depend on the solvent composition.

Hence, although Eqs. (5.76) to (5.80) point the way to deal *in principle* with the preferential solvation of large solute molecules by means of the inverse Kirkwood–Buff integral method, no actual demonstration of its use has been reported so far.

It is suggested that the random contact point method of Bloemendal et al. [109–111] can be used for the description of the interactions between functional groups of the large molecule. In particular, if the large molecule is a protein or a hydrocarbon (e.g., polypropylene), ester (polymethacrylate) or amidic (e.g., nylon) polymer, the interactions of the functional groups with solvent molecules can be estimated in this manner. This method was developed for the description of thermodynamic quantities Y, such as the Gibbs energy or enthalpy, of solutes (including two unlike solutes) in single solvents but should be extendable to single solutes (model compounds as indicated previously) in solvent mixtures. Because

TABLE 5.10 Group Surface Areas S_p and Enthalpic ($y_{pq} = h_{pq}$) and Gibbs Energetic ($y_{pq} = g_{pq}$) Interaction Parameter Differences $\Delta y_{pq} = y_{pq} - 0.5(y_{pp} + y_{qq})$, in kJ mol^{-1} for the Random Contact Point Model

Group p	S_p	Δh_{pq} group q			Δg_{pq} group q		
		H$_2$O	amide	–OH	H$_2$O	amide	–OH
–CH$_3$	1.57	–3.2	7.4	4.5	2.7	9.8	8.2
–CH$_2$–	1.00	–3.2	7.4	4.5	2.7	9.8	8.2
>CH–	0.42	–3.2	7.4	4.5	2.7	9.8	8.2
>C<	0.00						
>CO	1.19						
–CONH$_2$	2.47	5.5		1.4	2.6		2.0
–CONH–	1.92	5.5		1.4	2.6		2.0
–CON<	1.36	5.5		1.4	2.6		2.0
–OH	1.08	3.4	1.4		1.5	2.0	
–NH$_2$	1.28						
–NH–	0.73						
>N–	0.17						
H$_2$O	1.67						

The following additional values of h_{pq}/kJ mol^{-1} are available: p = alkyl (–CH$_3$, –CH$_2$–, and >CH–) and $q = p$, –7.6; q = –CON<, –23.8; q = –CONH–, –29.8; p = –CON< and $q = p$, –54.4; q = –CONH–, –44.8; $p = q$ = –CONH–, –44.4.
Source: Refs. 109–111.

the solutions of large molecules tend to be dilute, the group interactions derived from expressions for Y in terms of virial coefficient expansion of molalities (moles solute per kg of solvent) are appropriate. The representation of the measured excess quantity relative to infinite dilution, $Y^{E(\infty)} = \sum_n B_n^Y m^n$, yielded Gibbs energy and enthalpic group contributions that were found to be transferable from one solvent to another. The method also requires the use of the van der Waals surface areas S (relative to that of $-CH_2-$) of the groups. These quantities are shown in Table 5.10.

The applicable expression for the thermodynamic function of solvation (to dilutions where solute–solute interactions can be ignored) is [109–111]

$$\Delta_{solv} Y_u^\circ = \sum_p \sum_q [(2F_{u,p}F_{v,q} \sum F_v$$
$$- F_{v,p}F_{v,q} \sum F_u) y_{pq}]/2(\sum F_v)^2 \qquad (5.81)$$

where subscripts p and q denote groups, subscript u denotes the solute(s), subscript v denotes the solvent(s), $F_{i,j} = n_j S_j$ in a molecule of $i = u$ or v, n_j is the number of j groups (of kinds p or q), each having a relative van der Waals surface area S_j, $\sum F_i = \sum_j F_{i,j}$, and y_{pq} is the molar contribution of a p–q interaction to the property Y. As mentioned, this and related expressions have so far been applied only to single solutes or pairs of unlike solutes in single solvents, but the transferability of the group contributions encourages one to extend this to solvent mixtures containing the groups under discussion. The reported [109–111] quantities were mainly the differences $\Delta y_{pq} = y_{pq} - 0.5(y_{pp} + y_{qq})$ and also a few h_{pq} values; see Table 5.10. This approach, however, has so far not been tested.

REFERENCES

1. FJC Rossotti, H Rossotti. The Determination of Stability Constants. McGraw-Hill, New York, 1961.
2. MT Beck. Chemistry of Complex Equilibria. Van Nostrand, London, 1970.
3. P Chatterjee, S Bagchi. J Chem Soc Faraday Trans 86:1785, 1990.
4. P Chatterjee, S Bagchi. J Chem Soc Faraday Trans 87:587, 1991.
5. P Chatterjee, K Medda, S Bagchi. J Solution Chem 20:249, 1991.
6. P Chatterjee, S Bagchi. J Phys Chem 95:3311, 1991.
7. AK Laha, D Banerjee, S Bagchi. Indian J Chem A34:335, 1995.
8. AK Laha, PK Das, D Banerjee, S Bagchi. J Chem Soc Faraday Trans 92:1499, 1996.
9. T Fayed, SEH Etaiw. Spectrochim Acta A54:1909, 1998.

10. JA Soroka, KB Soroka. J Phys Org Chem 10:647, 1997.
11. DJ Phillips, JF Brennecke. Ind Eng Chem Res 32:943, 1993.
12. A Taha, AAT Ramadan, MA El-Behari, AI Ismail, MM Mahmoud. New J Chem 25:1306, 2001.
13. WE Acree, DC Wilkins, SA Tucker. Appl Spectrosc 47:1171, 1993.
14. WE Acree, SA Tucker, DC Wilkins. J Phys Chem 97:11199, 1993.
15. WE Acree, DC Wilkins, SA Tucker, JN Griffin, JR Powell. J Phys Chem 98:2537, 1994.
16. WE Acree, JR Powell. Phys Chem Liq 30:63, 1995.
17. WE Acree, SA Tucker, DC Wilkins, JN Griffin. Phys Chem Liq 30:79, 1995.
18. P Suppan. J Chem Soc Faraday Trans 1 83:495, 1987.
19. C Lerf, P Suppan. J Chem Soc Faraday Trans 1 88:963, 1992.
20. MCR Symons. Pure Appl Chem 58:1121, 1986.
21. G Eaton, MCR Symons. J Chem Soc Faraday Trans 1 84:3459, 1988.
22. G Eaton, MCR Symons, PP Rastogi, C O'Duinn, EW Waghorne. J Chem Soc Faraday Trans 1 88:1137, 1992.
23. G Eaton, L Harris, K Patel, MCR Symons. J Chem Soc Faraday Trans 1 88:3527 (1992).
24. M Meade, WE Waghorne, MCR Symons. J Chem Soc Faraday Trans 92:4395, 1996.
25. GR Bebahani, D Dunnion, P Falvey, K Hickey, M Meade, Y McCarthy, MCR Symons, WE Waghorne. J Solution Chem 29:521, 2000.
26. OB Nagy, M wa Muanda, JB Nagy. J Chem Soc Faraday Trans 1 74:2210, 1978.
27. JH Purnell, JM Vargas de Andrade. J Am Chem Soc 97:3585, 3590, 1975.
28. RJ Laub, JH Purnell. J Am Chem Soc 98:30, 35, 1976.
29. M wa Muanda, JB Nagy, OB Nagy. Tetrahedron Lett 38:3421, 1974.
30. B Parbhoo, OB Nagy. J Chem Soc Faraday Trans 1 82:1789, 1986.
31. WE Acree, AI Zvaigzne, SA Tucker. J Chem Soc Faraday Trans 86:307, 1990.
32. WE Acree Jr, GL Bertrand. J Phys Chem 81:1170, 1977; WE Acree Jr, GL Bertrand. J Phys Chem 83:2355, 1979; WE Acree Jr, GL Bertrand. J Pharm Sci 70:1033, 1981; WE Acree Jr, GL Bertrand. J Solution Chem 12:101, 1983.
33. WE Acree Jr, JH Ryttig. J Pharm Sci 71:201, 1982; WE Acree Jr, JH Ryttig. Int J Pharm 10:231, 1982.
34. WE Acree. Phys Chem Liq 22:107, 1990; JR Powell, MER McHale, ASM Kauppila, WE Acree, SW Campbell. J Solution Chem 25:1001, 1996.
35. MER McHale, JR Powell, ASM Kauppila, WE Acree, PL Huyskens. J Solution Chem 25:1089, 1996.
36. PL Huyskens, GG Siegel. Bull Soc Chim Belg 97:821, 1988.
37. CB Kretschmer, R Wiebe. J Chem Phys 22:1697, 1954.
38. WE Acree. Phys Chem Liq 22:107, 1990.

39. A Ben-Naim. Cell Biophys 12:255, 1988; A Ben-Naim. J Phys Chem 93:3809, 1989.
40. D Bannerjee, AK Laha, P Chatterjee, S Bagchi. J Solution Chem 24:301, 1995.
41. Y Marcus. Aust J Chem 36:1718, 1983.
42. EA Guggenheim. Proc R Soc (Lona) A169:134, 1938.
43. K Hagemark. J Phys Chem 72:2316, 1968.
44. Y Marcus. Pure Appl Chem 62:2069, 1990; Y Marcus. J Chem Soc Faraday Trans l 84:1465, 1988.
45. SH Yalkowski, GL Amidon, G Zografi, GL Flynn. J Pharm Sci 64:48, 1974.
46. D Khossravi, KA Connors. J Pharm Sci 81:371, 1992.
47. MA Postigo, M Katz. J Solution Chem 16:1015, 1987.
48. SM Leshchev. In Y Marcus, AK SenGupta, eds. Ion Exchange and Solvent Extraction 15. Marcel Dekker, New York, 2001.
49. GA Krestov, VP Korolyov, DV Batov. Thermochim Acta 169:69, 1990.
50. L Bernazzani, S Cabani, G Conti, V Mollica. J Chem Soc Faraday Trans 91:649, 1995.
51. BG Cox, EW Waghorne. Chem Soc Rev 9:381, 1980.
52. E DeValera, D Feakins, EW Waghorne. J Chem Soc Faraday Trans l 79:1061, 1983.
53. BG Cox, EW Waghorne. J Chem Soc Faraday Trans l 80:1267, 1984.
54. W Balk, G Somsen. J Phys Chem 89:5093, 1985; G Somsen. NATO ASI Ser C 119:425, 1984.
55. C DeVisser, G Somsen. J Phys Chem 78:1719, 1974.
56. Y Marcus. Pure Appl Chem 58:1721, 1986.
57. C Kalidas, GT Hefter, Y Marcus. Chem Rev 100:819, 2000.
58. GT Hefter, Y Marcus, WE Waghorne. Chem Rev 102:in press (2002).
59. H Ohtaki. Bull Chem Soc Jpn 42:1573, 1969.
60. H Ohtaki, N Tanaka. J Phys Chem 75:90, 1971.
61. H Ohtaki, M Maeda. Bull Chem Soc Jpn 46:2052, 1973.
62. H Ohtaki. Chem Lett (Chem Soc Jpn) 1973:439.
63. J Barbosa, V SanzNebot. J Chem Soc Faraday Trans 90:3287, 1994; J Barbosa, D Barron, R Berges, V SanzNebot, I Toro. J Chem Soc Faraday Trans 93:1915, 1997; D Barron, S Buti, M Ruiz. J Barbosa. Phys Chem Chem Phys 1:295, 1999; D Barron, S Buti, M Ruiz. J Barbosa. Polyhedron 18:3281, 1999.
64. DG Hall. J Chem Soc Faraday Trans 2 68:25, 1972.
65. H Schneider, H Strelow. Z Phys Chem (Frankfurt) 49:44, 1966.
66. A-KS Labban, Y Marcus. J Solution Chem 20:221, 1991.
67. A-KS Labban, Y Marcus. J Solution Chem 26:1 1997.
68. BG Cox, JA Parker, EW Waghorne. J Phys Chem 78:1731, 1974.
69. AK Covington, KE Newman. Pure Appl Chem 51:2041, 1979.
70. AK Covington, M Dunn. J Chem Soc Faraday Trans l 85:2827, 2835, 1989.

71. K Remerle, JBFN Engberts. J Phys Chem 87:5449, 1983.
72. AK Covington, KE Newman. J Chem Soc Faraday Trans 1 84:1393, 1988.
73. R Jellema, J Bulthuis, G van der Zwan, G Somsen. J Chem Soc Faraday Trans 92:2563, 1996.
74. R Jellema, J Bulthuis, G van der Zwan, G Somsen. J Chem Soc Faraday Trans 92:2569, 1996.
75. G Scatchard. J Chem Phys 9:34, 1941.
76. J Padova. Electrochim Acta 12:1227, 1967; J Padova. J Phys Chem 72:796, 1968.
77. H. Schneider. In J Coetzee, CD Ritchie, eds. Solute-Solvent Interactions. Marcel Dekker, New York, 1969, p 301.
78. RL Benoit, C Buisson. Inorg Chim Acta 7:256, 1973.
79. F Tanaka, Y Kawasaki, S Yamashita. J Chem Soc Faraday Trans 1 84:1083, 1988.
80. H Kleeberg. J Mol Struct 237:187, 1990.
81. VA Sajeevkumar, S Singh. J Mol Struct 382:101, 1996.
82. DS Gill, L Rodehüser, J-J Delpuech. J Chem Soc Faraday Trans 86:2847, 1990.
83. A Fratiello, DC Douglas. J Chem Phys 39:2017, 1963.
84. A Fratiello, EG Christie. Trans Faraday Soc 61:306, 1965.
85. A Fratiello, D Miller. J Chem Phys 42:796, 1965.
86. A Fratiello, DP Miller. Mol Phys 11:37, 1966.
87. JH Swinehart, H Taube. J Chem Phys 37:1579, 1962.
88. CK Finter, HG Hertz. Z Phys Chem NF 148:75, 1986; CK Finter, HG Hertz. J Chem Soc Faraday Trans 1 84:2735, 1988.
89. M Holz. J Chem Soc Faraday Trans 1 74:644, 1978.
90. M Holz, CK Rau. J Chem Soc Faraday Trans 1 78:1899, 1982; BM Braun, M Holz. J Solution Chem 12:685, 1983; M Holz, A Sacco. Mol Phys 54:149, 1985.
91. A Clausen, AA El-Harakany, H Schneider. Ber Bunsenges Phys Chem 77:994, 1973.
92. HH Füldner, DH Devia, H Strelow. Ber Bunsenges Phys Chem 82:499, 1978.
93. HH Füldner, H Strelow. Ber Bunsenges Phys Chem 86:68, 1982.
94. K Miura, H Matsuda, S Kikuchi, H Fukui. J Chem Soc Faraday Trans 87:837, 1991.
95. AJ Parker. Aust J Chem 16:585, 1963.
96. AJ Parker, U Mayer, R Schmidt, V Gutmann. J Org Chem 43:1843, 1978.
97. MJ Blandamer, J Burgess, JBFN Engberts. Chem Soc Rev 14:237, 1986; MJ Blandamer, NL Blundell, J Burgess, HJ Cowles, JBFN Engberts, IM Horn, P Warrick Jr. J Am Chem Soc 112:6854, 1990; MJ Blandamer, NL Blundell, J Burgess, HJ Cowles, IM Horn. J Chem Soc Faraday Trans 86:277, 1990.

98. H Strelow, H-M Koepp. Z Elektrochem 62:373, 1958.
99. AV Varghese, C Kalidas. Z Naturforsch A46:703, 1990; TK Sree Kumar, C Kalidas. J Indian Chem Soc 66:615, 1989; S Subramanian, C Kalidas. Fluid Phase Equil 32:205, 1987.
100. T Radnai, I Bako, G Palinkas. Acta Chim Hung 132:159, 1995.
101. T Takamuku, A Yamaguchi, D Matsuo, M Tabata, M Kumamoto, J Nishimoto, K Yoshida, T Yamaguchi, M Nagao, T Otomo, T Adachi. J Phys Chem B 105:6236, 2001.
102. S Galera, JM LLuch, A Oliva, J Bertran. J Chem Soc Faraday Trans 88:3537, 1992.
103. YM Kessler, II Vaisman, MG Kiselev, YP Puhovski. Acta Chim Hung 129:787, 1992.
104. SY Noskov, MG Kiselev, AM Kolker. Zh Fiz Khim 75:380, 2001.
105. YP Puhovski, BM Rode. J Mol Liq 91:149, 2001.
106. A Ben-Naim. Pure Appl Chem 62:25, 1990.
107. A Ben-Naim. Statistical Mechanics for Chemists and Biochemists. Plenum Press, New York, 1992, pp 436–440.
108. RA Pierotti. Chem Rev 76:717, 1976.
109. M Bloemendal, Y Marcus. AIChE J 33:1800, 1987.
110. M Bloemendal, Y Marcus, M Booij, R Hofstee, G Somsen. J Solution Chem 17:15, 1988.
111. M Bloemendal, Y Marcus. J Solution Chem 18:437, 1989.

6

Multicomponent Solvent Mixtures

Although binary solvent mixtures find wide applications in both the laboratory and industry, many actual liquid phases employed as solvents have more than two components. These can be independent of each other or, in some cases, dependent on the system as a whole.

For instance, consider a solvent extraction system where the organic phase is composed of an extractant, e.g., tributyl phosphate (TBP), and a diluent, e.g., dodecane, opposed to water as the aqueous phase. This aqueous phase may also be a solution of a salting-out agent, e.g., aluminum nitrate, that is not extracted itself. The organic phase in such a case will *not* be a *binary* mixture of TBP and dodecane but will be a *ternary* solvent mixture instead, also containing water. The concentration of the water depends on that of the salting-out agent in the aqueous phase, if present, whether solutes that distribute between the two phases are present at low concentrations or not. In the absence of a salting-out agent the concentration of water in the organic phase is determined by (the temperature and) the relative concentrations of the TBP and the dodecane.

Other multicomponent solvent mixtures may consist of natural mixtures (e.g., petroleum-derived hydrocarbon solvents, such as

239

kerosene) or mixtures obtained in a synthetic process, e.g., in hydration of olefins (to produce alkanols, such as amyl alcohol), where efforts to isolate pure products are not cost-efficient. On the other hand, formulations of multicomponent solvent mixtures are often made intentionally to obtain the most desirable properties for the envisaged application. The physical and chemical properties of such multicomponent solvent mixtures are therefore of importance for the understanding of their behavior in applications and for the selection of their optimal compositions, where choices can be made.

The compositions of multicomponent solvent mixtures A, B, C, ... are usually specified in terms of mole fractions, x_A, x_B, x_C, ... with $\sum x_i = 1$; mass (weight) fractions, w_A, w_B, w_C, ... with $\sum w_i = 1$; or volume fractions, ϕ_A, ϕ_B, ϕ_C, ... with $\sum \phi_i = 1$. For some purposes the molar concentrations, c_A, c_B, c_C, ... or the number densities, ρ_A, ρ_B, ρ_C, ... may find use. However, the molal scale is very rarely used because it is seldom possible to select a certain component of the liquid mixture as being *the* solvent, per kg of which the other concentrations are specified.

In the following, ternary solvent mixtures are discussed, albeit only briefly, because extension to higher multicomponent mixtures is self-evident.

6.1 THERMODYNAMIC AND PHYSICAL PROPERTIES OF MULTICOMPONENT MIXTURES

The thermodynamic properties of ternary solvent mixtures, designated by the generalized symbol Y, are in general not linear with the composition. That is, the excess function Y^E in

$$Y = x_A Y_A + x_B Y_B + x_C Y_C + Y^E \tag{6.1}$$

is not zero. On the other hand, the excess function can in many cases be related to the corresponding excess functions in the binary partial mixtures:

$$Y^E = x_A x_B Y^E_{AB} + x_A x_C Y^E_{AC} + x_B x_C Y^E_{BC} + x_A x_B x_C Y^E_{ABC} \tag{6.2}$$

where the ternary function Y^E_{ABC} can often—but not in all cases—be neglected. Bertrand et al. [1] reviewed the methods available at the time for the evaluation of $Y^E = G^E$, H^E, S^E, and V^E of ternary systems and suggested a form suitable for computer parametrization. Conti et al. [2] studied all the pertinent excess thermodynamic functions, G^E, H^E, S^E, C_p^E, and V^E, for the system comprising ethanol, tetrahydrofuran (THF), and cyclohexane. In this ternary system all five excess functions are positive, although some of the excess functions for the binary subsystems are negative, e.g., S^E for the ethanol + cyclohexane mixtures. Addition of the THF to this binary mixture eventually reverses the sign of S^E.

Many of the expressions written for the thermodynamic properties of binary solvent mixtures (Section 2.3) can be extended to deal with ternary mixtures. For instance, the Margules two-suffix expression for the excess Gibbs energy:

$$G^E = RT \sum_i \sum_j A_{ij} x_i x_j \quad (i, j = A, B, \text{ and } C, A_{ij} = A_{ji}) \tag{6.3}$$

setting $Y^E_{ABC} = G^E_{ABC} = 0$ in Eq. (6.2), can be employed to describe vapor – vapor equilibria (VLE). The corresponding Redlich–Kister expression is

$$\begin{aligned} G^E = RT[&x_A x_B G^E_{AB} + x_A x_C G^E_{AC} + x_B x_C G^E_{BC}] \\ &+ x_A x_B x_C [A_0 + A_{1AB}(x_A - x_B) \\ &+ A_{1AC}(x_A - x_C) + A_{1BC}(x_B - x_C) + \cdots] \end{aligned} \tag{6.4}$$

where the additional terms contain higher powers of the differences in the mole fractions. These expressions can serve for the mathematical fitting of the data but not for the prediction of values outside the ranges where data have been determined.

The prediction of ternary quantities from the binary ones is possible when several of the expressions used for binary mixtures, Section 2.3, are employed. Wilson's expression has been modified by Tsuboka and Katayama [3] for ternary systems to read

$$G^E = -RT \sum_i x_i \left[\ln \sum_j x_j \Lambda_{ij} + \ln \sum_j x_j (V_i / V_j) \right] \tag{6.5}$$

$$(i, j = A, B, \text{ and } C)$$

Here only the three binary parameters Λ_{AB}, Λ_{AC}, and Λ_{BC} are required for the prediction of the ternary excess Gibbs energy.

The nonrandom two-liquid (NRTL) expression of Renon and Prausnitz becomes for ternary systems [4]:

$$G^E = RT \sum_i x_i \sum_j \tau_{ji} G_{ji} x_j / \sum_k G_{ki} x_k \quad (i, j, k = A, B, \text{ and } C)$$

$$\tag{6.6}$$

where $\tau_{ji} = (g_{ji} - g_{ii})/RT$, $G_{ji} = \exp(-\alpha_{ji}\tau_{ji})$, $g_{ji} = g_{ij}$, $\tau_{ji} \neq \tau_{ij}$, $G_{ji} \neq G_{ij}$, and $\alpha_{ji} = \alpha_{ij}$ ($= \alpha$ if a common value for this parameter is chosen). Six binary parameters—τ_{AB}, τ_{BA}, τ_{AC}, τ_{CA}, τ_{BC}, and τ_{CB}—are required for the use of Eq. (6.6) (if a common α value is employed), but no ternary parameter. Renon and Prausnitz [4] applied their expression to the vapor pressure and vapor compositions of systems comprising three hydrocarbons, two hydrocarbons + a polar solvent, two hydrocarbons + a hydrogen-bonding solvent, and three polar solvents, including one or two hydrogen-bonding ones, with excellent results.

Abrams and Prausnitz [5] showed that the UNIQUAC expression can be similarly extended from binary to ternary liquid mixtures. The combinatorial term, Eq. (2.27) of Section 2.3, now reads

$$G_{comb}^{E}/RT = \sum_i x_i \ln(\phi_i/x_i) + (Z/2) \sum_i q_i x_i \ln(\theta_i/\phi_i)$$

$$(i = A, \ B, \ \text{and} \ C) \tag{6.7}$$

and depends only on the surface area fraction θ_i and volume fraction ϕ_i of the pure components. The interaction part, Eq. (2.28) of Section 2.3, now reads

$$G_{inter}^{E}/RT = -\sum_i q_i x_i \ln\left[\sum_j \theta_j \tau_{ji}\right] \quad (i,j = A, B, \text{and C}) \tag{6.8}$$

where $\tau_{ji} = \exp[-(u_{ji} - u_{ii})/RT]$ and $\tau_{ji} \neq \tau_{ij}$. Again, six interaction parameters τ are required. Abrams and Prausnitz [5] applied their expressions to ternary systems involving nonpolar, polar, and hydrogen-bonding solvents with good success. In particular, they showed that their equations are capable of describing the ternary phase diagram when liquid–liquid immiscibility takes place. Conti et al. [2] showed that the UNIQUAC method is able to predict G^E and H^E for the ternary ethanol + THF + cyclohexane system better than other methods. The use of the NRTL and the UNIQUAC expressions for such cases was further explored by Sørensen et al. [6]. Nagata and Kawamura [7,8] employed the UNIQUAC expressions for the ternary system methanol + acetonitrile + chlorobenzene that contained the self-associating methanol. They used an extension of the UNIQUAC expressions to allow for specific interactions. Hence, calculations of the vapor pressure and vapor composition also included the self-association equilibrium constant of methanol and the interaction constant between methanol and each of the other components [7,8].

Not only can excess Gibbs energies and VLE data of ternary solvent mixtures be described by, essentially, Eq. (6.2) without the need for a ternary parameter, but heats of mixing in ternary systems have been so described equally successfully. Bertrand et al. [1], Prchal et al. [9], and Pando et al. [10] reviewed the methods proposed at the time for the prediction of ternary H^E from binary data. The latter authors applied the insight gained to 42 ternary systems. For the binary subsystems, the addition of a factor in the denominator of the Redlich–Kister expression:

$$H^E_{ij} = x_i x_j \left[\sum A_k (x_i - x_j)^k \right] \bigg/ B(x_i - x_j) \qquad (6.9)$$

permits the fitting of very asymmetrical excess enthalpy curves, including those that change from endo- to exothermic, better than the commonly used expression with $B = 0$, restricted to slightly asymmetrical curves. Equation (6.2) is then used for the prediction of the ternary H^E, setting the ternary term in $H^E_{ABC} = 0$. An alternative expression suggested is:

$$H^E = \sum \sum_k A_k x_i x_j / (x_i / B_k + x_j B_k)^2 \qquad (i,j = A, B, \text{ and } C)$$

$$(6.10)$$

(the coefficients A and B are different for the various pairs of i and j, and the index k may go up to 3). The most successful predictive expression for ternary systems was one where component A was selected as the component most chemically different from the other two (i.e., H^E_{BC} is the lowest in the absolute sense), using the expression

$$H^E = x_A x_B \left[\sum A_{k(AB)} (2x_A - 1)^k \right] \bigg/ B_{(AB)} (2x_A - 1)$$

$$+ x_A x_C \left[\sum A_{k(AC)} (2x_A - 1)^k \right] \bigg/ B_{(AC)} (2x_A - 1) \qquad (6.11)$$

$$+ x_B x_C \left[\sum A_{k(BC)} (x_B - x_C)^k \right] \bigg/ B_{(BC)} (x_B - x_C)$$

Thus, when two alcohols and an alkane were involved, the alkane should be selected as component A, but when an alcohol mixed with two hydrocarbons or a hydrocarbon and an alkyl halide are considered, the alcohol should be selected as component A.

Nagata et al. [11–15] dealt with several ternary systems, containing no, one, and two self-associating components (these were methanol and/or ethanol), using an extension of the NRTL approach that included self- and mutual binary interactions characterized by temperature-dependent equilibrium constants and enthalpies of interaction. The expressions used are quite formidable and are not reproduced here. On the other hand, specific ternary terms have been found to be required in other cases. These include those involving pyridine, β-picoline, and benzene or c-hexane and benzene, 1,2-dichloroethane, and toluene or o- or p-xylene, studied by Singh and Sharma [16]. Similarly, systems comprising c-hexane, tetrachloromethane, and benzene or acetone and c-hexane, chloroform, and acetone, studied by Lark et al. [17], and 1,4-dioxane, tetrachloromethane, and a third component, studied by Acevedo et al. [18], require ternary terms.

For the prediction of the ternary excess heat of mixing from the binary ones, the expression

$$H^E = (x_A + x_B)^2 H^E_{AB} + (x_A + x_C)^2 H^E_{AC} + (x_B + x_C)^2 H^E_{BC} \qquad (6.12)$$

[19] without an explicit ternary term, H^E_{ABC}, can be used, but the fits are inferior to those achieved when a ternary term is added. This could be done, e.g., according to the fluid mixture theory of Sanchez and Lacombe [16,20]. Other thermodynamic functions of ternary systems, such as excess volume and heat capacities, should be describable in similar terms. A further discussion of the excess enthalpy in ternary systems was published by Korolev [20a], involving preferential solvation and the extended solvent coordination model.

Fewer expressions have been suggested for the prediction of the excess molar volume of ternary solvent mixtures. A reason for this could be the relative ease of direct measurement of the density of such mixtures, obviating the necessity for its calculation. The density is, after all, the quantity mostly relevant for the engineering design where such mixtures are employed. Rastogi et al. [21] suggested the following expression:

$$V^E = 0.5[(x_A + x_B)V^E_{AB} + (x_A + x_C)V^E_{AC} + (x_B + x_C)V^E_{BC}] \qquad (6.13)$$

where the binary excess volumes pertain to the compositions $x_A/(x_A + x_B)$ for V^E_{AB}, etc. However, with this expression when x_C approaches infinite dilution, $V^{\not E}$ does not approach V^E_{AB} as it should. Other expressions were suggested by Bertrand et al. [1] and by Dahiya et al. [22]. The latter authors showed that their expression:

$$V^E = \sum_{ij} x_i x_j [v_{0ij} + v_{1ij}(x_i - x_j) + v_{2ij}(x_i - x_j)^2]$$

$$+ x_A x_B x_C [v_{0ABC} + v_{1ABC}(x_B - x_C)x_A + v_{2ABC}(x_B - x_C)^2 x_A^2]$$

$$(i,j = A,B; \ A,C; \ \text{and BC}) \tag{6.14}$$

is also applicable to the excess enthalpy, H^E, with appropriate parameters h. It described well excess volumes (and enthalpies, if parameters h are employed) of four ternary aromatic solvent mixtures with the required nine binary as well as three ternary parameters.

Other physical properties of ternary solvent systems, such as the viscosity, have also been related to the properties of the constituent binaries. The kinematic viscosity $v_{Mi} = \eta_i M_i / d_i$ for each component in a mixture was supposed by Arroyo [23] to be linearly additive to give the dynamic viscosity, η:

$$\eta = \left(\sum_i x_i v_{Mi} \right) d \bigg/ \sum_i x_i M_i \tag{6.15}$$

This should hold for mixtures of nonpolar solvents, where no association occurred. When association takes place, however, the presence of the clusters formed should be taken into account by noting the stoichiometric ratio in the binary mixture where the maximal viscosity was found and ascribing to the cluster this maximal value. The composition was then taken to be that of the cluster plus the component that was in excess, and Eq. (6.15) was applied. For ternary mixtures the absence of ternary clusters was assumed and a hierarchical order of binary clustering was established for the application of Eq. (6.15). This approach was applied to various mixtures containing water, the lower alkanols, and acetone. A more elaborate expression, based on the four-body interaction model of McAllister [24], was proposed by Noda et al. [25] and was applied to the water + methanol + acetone mixture, also studied by Arroyo [23], in good agreement.

6.2 PREFERENTIAL SOLVATION IN TERNARY SOLVENT MIXTURES

The preferential solvation characteristics of ternary liquid systems can be obtained by the application of the Kirkwood–Buff integrals method without requiring the standard molar transfer Gibbs energies of a solute. Rather, the composition dependence of the activity coefficients or the excess Gibbs energies in these systems is used in expressions analogous to Eqs. (4.25) and (4.26), Section 4.2. Matteoli and Lepori [26,27] studied ternary systems and focused their attention on the Kirkwood–Buff integrals and in particular on the corrected values relative to the ideal integrals, $\Delta G = G - G^{id}$, as was done in the case of binary systems (see Section 4.3). Furthermore, they did not deal exclusively with infinite dilutions of a solute in the binary solvent systems, Section 5.2.2, but considered substantial concentrations of all three components: A, B, and C. Thus, they, discussed not only the interactions of, say, component C with the solvents A and B but also the effects that C has on the A–B interactions. The preferential solvation coefficient, psc, that they calculated was defined analogously with Eq. (4.30). However, for the coefficient of, say, component C it was divided by $x_A x_B$, so as not to lose information at infinite dilution of either A or B, as

$$psc^o_{C(AB)} = \Delta G_{AC} - \Delta G_{BC} \qquad (6.16)$$

There are altogether six Kirkwood–Buff integrals, corrected for volume discrepancies, $\Delta G_{ij} = G_{ij} - G_{ij}^{id}$, (see Section 4.3) for two-body interactions. Matteoli and Lepori [26,27] examined several ternary liquid systems with up to $x_C = 0.3$ in mixtures of A and B. They studied the interactions taking place in mixtures of water (A) with 1-propanol (B) in the presence of amides (C): urea at 298 K on the one hand and N,N-dimethylformamide (DMF) at 313 K on the other. In the presence of urea, the positive self-interaction (corrected) Kirkwood–Buff integrals ΔG_{AA}, ΔG_{BB}, and ΔG_{CC} as well as the mutual interaction integral ΔG_{AC} are enhanced, but the negative ΔG_{AB} and ΔG_{BC} become even more negative. On the contrary, in the presence of DMF, ΔG_{AA} and ΔG_{BB} strongly decreased and ΔG_{AB} was strongly raised, practically vanishing at $x_C = 0.3$. The integrals ΔG_{AC} and ΔG_{BC} are rather small for C = DMF. These opposing types of behavior are explained by the strong affinity between urea and water, leading to enhanced hydrophobic interactions between the propanol molecules (urea is a known good salting-out agent), but the effect of the less hydrophilic DMF is to loosen the self-interactions of the propanol due to steric hindrance.

In mixtures of water (A) and 1,2-ethanediol (EG) (B) containing acetonitrile (C) or ethanol (C), both at 323 K, the strong mutual affinity of water and EG is again manifest in the corresponding Kirkwood–Buff integrals [26,27]. As far as the interactions of acetonitrile or ethanol are concerned, however, the A + B mixture can be considered as a single solvent. The self-association of the acetonitrile is enhanced as is that of the ethanol, although to a lesser extent. In the ternary system involving a hydrogen bond accepting solvent dioxane (A), a mild hydrogen bond donating solvent chloroform (B), and a strong amphiprotic solvent ethanol (C), the interactions that are expected are indeed manifested by the Kirkwood–Buff integrals. The self-association of the ethanol is lower in the presence of dioxane than of chloroform, and self-repulsion of dioxane is observed in its dilute solutions in the other two components, confirming association between dioxane and the hydrogen bond donating solvents.

Zielkiewicz [28] studied N-methylformamide (NMF) (C) in mixtures of water (A) and either methanol (B) or ethanol (B) at 313 K. In these systems, as for the DMF + water + alcohol systems studied previously (see earlier), the preferential solvation in amide-rich mixtures is small and is almost entirely due to the size differences of the component substances. In alcohol-rich mixtures, however, some preferential hydration of the amide is observed. Later, N-methylacetamide (NMA) was added to the study [29], and it was seen that NMF does differ from both NMA and DMF, which behave similarly, in aqueous alcohol mixtures. Each of the latter two has two hydrophobic methyl groups, whereas NMF has only one. The deviations of the local mole fractions from the bulk ones are only a few percent in all cases, however. These results were confirmed by molecular dynamics calculations carried out at $x_{\text{amide}} = 0.518$.

Once the local mole fractions in a ternary solvent system have been established by, say, the Kirkwood–Buff integral method, this information can be used in order to explain the behavior of the system from other aspects. For instance, the partial molar enthalpies of the components can be related to the local mole fractions; hence the interactions taking place in the system could be elucidated by Matteoli [30]. If no preferential solvation of the solvent C in mixtures of solvents A and B takes place, $x^{\text{L}}_{\text{C(AB)}} - x_{\text{C}} \approx 0$, then its partial molar enthalpy, H_{C}, changes linearly between its values in the neat solvents, $H_{\text{C(A)}}$ and $H_{\text{C(B)}}$, when, say, x_{A} changes. When, however, the excess local mole fractions near C are appreciable, $|x^{\text{L}}_{\text{C(AB)}} - x_{\text{C}}| \gg 0$, then these play a major role in the interactions and in the magnitude of H_{C}, provided one of the solvents A or B is capable of self-association. The deviations then obtained from the linear dependence of H_{C} on x_{A} are due to the partial molar enthalpies of transfer of the solvents A and B from the bulk to the solvation shell of C as well as to specific interactions of the components. These considerations are illustrated by the results obtained for the systems dioxane (C) in mixtures of ethanol and chloroform and for acetone (C) in mixtures of methanol and chloroform where strong preferential solvation of C takes place. They are valid, too, for ethanol (C) in mixtures of dioxane and chloroform and for methanol (C) in mixtures of chloroform and acetone, where preferential solvation of C does not play a major role.

A different approach was used by Wang et al. [31] to model the excess enthalpies in ternary solvent mixtures based on the preferential solvation concept, i.e., local mole fractions differing from the bulk ones. Only the first solvation shell around a molecule of kind i is considered, with a square potential well for the interactions of molecules of kind j with i. The molar value of the potential interaction energy of the i–j pair is e_{ji} at the bottom of the square well of width σ_{ji} (equal to the collision diameter of the molecules). The radial distribution function for the interactions of molecules of kind j with i within this first solvation shell is assumed to be

$$g_{ji}(r) = \exp(-\alpha e_{ji}/RT) \tag{6.17}$$

The factor α is a temperature-dependent fitting parameter. The resulting local mole fraction of j molecules in the first solvation shell around an i molecule is

$$x_{ji}^{L} = x_j \sigma_{ji}^3 \exp(-\alpha e_{ji}/RT) \Big/ \sum_k x_k \sigma_{ki}^3 \exp(-\alpha e_{ki}/RT) \tag{6.18}$$

(the original paper [31] incorrectly omitted the division sign before the summation sign in this expression). The index k in the summation in the denominator of Eq. (6.18) pertains to all the kinds of molecules in the multicomponent mixture, including i. The binary excess enthalpy of mixing is equated with the excess configurational internal energy in a quasi-lattice with the lattice parameter $Z = 10$ (as in UNIQUAC expressions) leading to

$$H^{E} = (Z/2)[x_A x_{BA}^{L} \Delta e_{BA} + x_B x_{AB}^{L} \Delta e_{AB}] \tag{6.19}$$

where $\Delta e_{BA} = e_{BA} - e_{AA}$, $\Delta e_{AB} = e_{AB} - e_{BB}$. The local mole fractions are given by

$$x_{BA}^{L} = x_B(\sigma_{BA}/\sigma_{AA})^3 \exp(-\alpha \Delta e_{BA}/RT)/$$
$$[x_A + x_B(\sigma_{BA}/\sigma_{AA})^3 \exp(-\alpha \Delta e_{BA}/RT)] \tag{6.20}$$

and similarly for x_{BA}^{L} with subscript B replacing A, subscript AB replacing BA, and BB replacing AA. Three parameters, α, Δe_{BA}, and Δe_{AB}, are required for fitting the excess enthalpy data, and this was done successfully for 29 binary mixtures involving polar and hydrogen-bonding as well as nonpolar solvents [31]. This approach was extended to ternary systems, where

$$H^E = (Z/2) \sum_i \sum_j x_i x_{ji}^L \Delta e_{ji} \qquad (i,j = \text{A, B; A, C; and B C})$$

$$(6.21)$$

with the local mole fractions

$$x_{ji}^L = x_j (\sigma_{ji}/\sigma_{ii})^3 \exp(-\alpha \Delta e_{ji}/RT) / \sum_k (\sigma_{ki}/\sigma_{ii})^3 \exp(-\alpha e_{ki}/RT)$$

$$(6.22)$$

Expressions (6.21) and (6.22) were successfully applied to 12 ternary solvent systems. No specific ternary term was required in these cases. It is disappointing, in a sense, that the expressions for the local mole fractions, Eqs. (6.20) and (6.22), were not used [31] for the exploration of the preferential solvation that took place in these systems.

6.3 SOLUTE SOLUBILITY IN TERNARY SOLVENT MIXTURES

The solubilities of solutes—not themselves solvents—in ternary solvent mixtures also merit consideration. A large amount of work in this direction was carried out by Acree and coworkers [32–35], who measured the solubilities of large molecules, such as anthracene, in ternary solvent mixtures and interpreted their results. The solubility of anthracene (S) in mixtures of two alkanes (A and B) and an alkanol (C), two alkanols (A and B) and an alkane (C) or an alcohol-ether (C), and dioxane (A) + heptane (B) and an alkanol (C) was interpreted in terms of the combined NIBS (see Section 5.2.2) and Redlich–Kister equation:

$$\ln x_s^{\text{sat}} = \sum_i x_i \ln x_{s(i)}^{\text{sat}} + \sum_{ij} x_i x_j \sum_k s_{k(ij)} (x_i - x_j)^k$$

$$(i,j = \text{A, B, and C}) \qquad (6.23)$$

where the mole fractions refer to the binary mixtures without the solute and $x_{S(i)}^{\text{sat}}$ is the solubility of S in pure solvent i. Up to three binary $s_{k(ij)}$ parameters (for $k = 0$, 1, and 2) are required for each of the three binary sub-systems ij studied, obtained from the solubilities measured in them.

In a recent publication by Jouyban-Gharmaleki et al. [36], the use of volume fractions ϕ_i of the solvents instead of the mole fractions x_i was suggested. Also, the use of the modified Wilson equation:

$$1 + \ln(x_S^{\text{sat}}/x_S^{\text{sat id}}) = \sum_{ijk} \phi_i(1 + \ln(x_{s(i)}^{\text{sat}}/x_S^{\text{sat id}})/$$

$$(\phi_i + \Lambda_{ij}\phi_j + \Lambda_{ik}\phi_k) \qquad (6.24)$$

was explored. Here, $x_S^{\text{sat id}}$ is the solvent-independent ideal solubility of the solute S, obtained from its heat and temperature of fusion: $\ln x_S^{\text{sat id}} = -\Delta H_S^F(T^F - T)/RT^F T$, and there are six $\Lambda_{ij} \neq \Lambda_{ji}$ parameters for i and j being pairs among A, B, and C. Other expressions for the estimation of the solubility of a solute in ternary solvent mixtures were also reviewed in this study.

REFERENCES

1. GL Bertrand, WE Acree Jr, TE Burchfield. J Solution Chem 12:327, 1983.
2. G Conti, P Gianni, L Lepori, E Matteoli. Pure Appl Chem 67:1849, 1995.
3. T Tsuboka, T Katayama. J Chem Eng Jpn 8:181, 1975.
4. H Renon, JM Prausnitz. AIChE J 14:135, 1968.
5. DS Abrams, JM Prausnitz. AIChE J 21:116, 1975.
6. JM Sørensen, T Magnussen, P Rasmussen, A Fredenslund. Fluid Phase Equil 3:47, 1979.
7. I Nagata, Y Kawamura. Chem Eng Sci 34:601, 1979.
8. I Nagata. J Chem Thermodyn 16:955, 1984.
9. M Prchal, V Dohnal, F Vesely. Coll Czech Chem Commun 48:1104, 1983.
10. C Pando, JAR Renuncio, JAG Calzon, JJ Christensen, RM Izatt. J Solution Chem 16:503, 1987.
11. I Nagata, K Tamura, S Tokuriki. Thermochim Acta 47:315, 1981.
12. I Nagata, K Tamura, S Tokuriki. Fluid Phase Equil 8:75, 1982.
13. I Nagata, K Tamura. J Chem Thermodyn 15:721, 1983; 16:975, 1984.
14. I Nagata, K Tamura. Fluid Phase Equil 15:67, 1983; 24:289, 1985.
15. I Nagata, K Tamura. Thermochim Acta 77:281, 1984; 57:331, 1982.
16. PS Singh, SP Sharma. Thermochim Acta 83:253, 1985.
17. BS Lark, S Kaur, S Singh. Thermochim Acta 105:219, 1986.
18. IL Acevedo, GC Pedrosa, M Katz, J Chem Eng Data 41:391, 1996.
19. WE Acree, Jr. Thermodynamic Properties of Non-Electrolyte Solutions. Academic Press, New York, 1984.
20. IC Sanchez, RH Lacombe. J Phys Chem 80:2352, 1976.
20a. VP Korolev. Russ J Gen Chem 70:1859, 2000.
21. RP Rastogi, J Nath, SS Das. J Chem Eng Data 22:249, 1977.

22. HP Dahiya, PP Singh, S Dagar. Fluid Phase Equil 43:341, 1988.
23. A Arroyo. Rev Roum Chim 35:523, 1990.
24. RA McAllister. AIChE J 6:427, 1960.
25. K Noda, M Ohashi, K Ishida. J Chem Eng Data 27:326, 1982.
26. E Matteoli, L Lepori. J Mol Liq 47:89, 1990.
27. E Matteoli, L Lepori. J Chem Soc Faraday Trans 91:431, 1995; correction: J Chem Soc Faraday Trans 91:1885, 1995.
28. J Zielkiewicz. J Chem Soc Faraday Trans 94:1713, 1998.
29. J Zielkiewicz. Phys Chem Chem Phys 2:2925, 2000.
30. E Matteoli. J Mol Liqs 79:101, 1999.
31. L Wang, S Shen, GC Benson, BC-Y Lu. Fluid Phase Equil 95:43, 1994.
32. TH Deng, WE Acree Jr. J Chem Eng Data 43:1059, 1998.
33. TH Deng, WE Acree Jr. J Chem Eng Data 44:544, 1999.
34. TH Deng, CE Hernandez, LE Roy, WE Acree, Jr. J Chem Thermodyn 31:205, 1999.
35. KJ Pribyla, MA Spurgin, I Chuca, WE Acree Jr. J Chem Eng Data 45:965, 2000.
36. A Jouyban-Gharmaleki, BJ Clark, WE Acree, Jr. Chem Pharm Bull 48:1866, 2000.

Solvents and Solvent Mixture Index

Subject Index